Collins **AQA GCSE**

Grade Booster

Physics

Dan Foulder

Contents

Introduction 3
Command Words 7

1. Energy 9

2. Electricity 23

3. Particle Model of Matter 40

4. Atomic Structure 46

5. Forces 59

6. Waves 82

7. Magnetism and Electromagnetism 96

8. Space Physics 107

Glossary 114
Formulae and Physics Equations Sheet 121
Periodic Table 123

Introduction

About this Book

This book has been designed to support your preparation for the AQA GCSE Physics (9–1) examination and help you achieve your best possible grade.

The AQA Physics specification is divided into eight topics. This Grade Booster book mirrors that approach and the worked questions are divided into the same topics as on the specification, so you can easily find questions that cover every topic you will study.

As you revise each topic in this book, you will find exam-style questions, model answers and supporting notes with tips and hints. You will also find guidance on what the examiner is looking for.

Questions that only cover content included in the Higher Tier exam are shown in this book with this symbol: **HT**.

You can visit the AQA website to download or view a copy of the GCSE Physics specification.

Terms in **bold** are defined in the Glossary at the back of the book.

At the end of each chapter, you are signposted to pages in *Collins GCSE AQA Physics Revision Guide* (ISBN 9780008160692) for more information on the topics covered. The same page references apply to the *Collins GCSE AQA Physics All-in-One Revision & Practice* (ISBN 9780008160739).

AQA GCSE Physics Exams

You will sit **two exams**, each of 1 hour 45 minutes duration.

In the final year of your GCSE course, your school will choose to enter you for either the Higher or Foundation Tier exam. If you are not sure which exam tier you have been entered for, talk to your physics teacher.

Introduction

The information in the table below is the same whichever tier you sit.

	Paper 1	Paper 2
Topics covered	1. Energy 2. Electricity 3. Particle model of matter 4. Atomic structure	5. Forces 6. Waves 7. Magnetism and electromagnetism 8. Space physics *Questions in Paper 2 may draw on an understanding of energy changes and transfers due to heating, mechanical and electrical work and the concept of energy conservation from Energy and Electricity.*
Exam marks	100	100
% of overall grade	50%	50%
Types of questions	Multiple choice Structured Closed short answer Open response	Multiple choice Structured Closed short answer Open response

Grading and Certification

The qualification will be graded on a nine-point scale: 1–9, where 9 is the highest grade.

If you are taking the Foundation Tier exam, you will be awarded a grade within the range of 1–5. If you fail to reach the minimum standard for grade 1, you will be recorded as U (unclassified) and will not receive a qualification certificate.

If you are taking the Higher Tier exam, you will be awarded a grade within the range of 4–9. If you are sitting the Higher Tier exam and you narrowly fail to achieve grade 4, you will be awarded a grade 3. If you fail to reach the minimum standard for the allowed grade 3, you will be recorded as U (unclassified) and will not receive a qualification certificate.

Introduction

Assessment Objectives

There are three assessment objectives (AOs) and the two exams will test these three different areas.

Assessment objective	Percentage of exam	Requirements
AO1 Demonstrate knowledge and understanding	40%	Demonstrate knowledge and understanding of: • scientific ideas • scientific techniques and procedures.
AO2 Apply knowledge and understanding	40%	Apply knowledge and understanding of: • scientific ideas • scientific enquiry • scientific techniques and procedures.
AO3 Analyse information and ideas	20%	Analyse information and ideas to: • interpret and evaluate • make judgements and draw conclusions • develop and improve experimental procedures.

Working Scientifically

The study of physics will involve a lot of facts, theories and explanations but also you will gain an understanding of the scientific process by thinking, discussing and reading about what scientists do. You will look at the contribution that some scientists have made to the world of physics and explore how they did this by thinking about ideas such as how models can be used to represent concepts in physics – a very important idea given that we can't easily see what is happening on an atomic level!

Required Practical Activities

Physics by its very nature is a practical subject. Many of the facts and ideas in this subject are derived from experiments, and theories can be proven or improved by experimental work.

The specification requires that you carry out practical work in ten specific areas, although you may well do more than this number of practical experiments during your course. Approximately 15% of marks in the exams will be based on the understanding that you have carried out these required practical activities. They will draw on the knowledge and understanding gained from having completed these practical activities.

There are examples of questions based on the required practical activities throughout this book. You can find further information about the required practical activities in the AQA specification.

Mathematical Requirements

Many topics in physics have a mathematical element to them. There are a number of formulae you need to learn and a number that will be given to you on the Physics Equations Sheet. A summary of all these formulae can be found on the Physics Equations Sheet on pages 121–122.

There are also many mathematical skills that you will need to be confident with, such as rearranging equations and calculating percentages. The questions in this book cover the mathematical skills required and you can find further information about the mathematical skills needed by looking at the AQA specification.

Success in Physics

Physics can sometimes be a difficult subject that students find quite intimidating! There are lots of facts to learn, theories to understand and explain, you are expected to apply your knowledge in different contexts AND do quite a lot of maths!

Practising questions such as those given in this book will help you to assess your knowledge and understanding. It will also help you learn how to answer exam questions successfully. However, for factual information and to aid your understanding of the subject, you can use this book in conjunction with the *Collins GCSE AQA Physics Revision Guide* (ISBN 9780008160692) or *Collins GCSE AQA Physics All-in-One Revision & Practice* (ISBN 9780008160739).

Command Words

Command words are specific words in questions that tell you what is expected in your answer. Command words can help you decide how to answer questions, how much detail to give and whether you are expected to simply recall information or give more explanation.

As you work through this book try to identify the command word in each question. If you are not sure what a question is asking you to do then refer back to this list of words and their explanations.

Command word	Explanation
Calculate	students should use numbers given in the question to work out the answer
Choose	select from a range of alternatives
Compare	this requires the student to describe the similarities and / or differences between things, not just write about one
Complete	answers should be written in the space provided, for example on a diagram, in spaces in a sentence, or in a table
Define	specify the meaning of something
Describe	students may be asked to recall some facts, events or processes in an accurate way
Design	set out how something will be done
Determine	use given data or information to obtain an answer
Draw	produce, or add to, a diagram
Estimate	assign an approximate value
Evaluate	students should use the information supplied, as well as their knowledge and understanding, to consider evidence for and against
Explain	make something clear, or state the reasons for something happening
Give	only a short answer is required, not an explanation or a description
Identify	name or otherwise characterise
Justify	use evidence from the information supplied to support an answer
Label	provide appropriate names on a diagram
Measure	find an item of data for a given quantity

Command Words

Name	only a short answer is required, not an explanation or a description; often it can be answered with a single word, phrase or sentence
Plan	write a method
Plot	mark on a graph using data given
Predict	give a plausible outcome
Show	provide structured evidence to reach a conclusion
Sketch	draw approximately
Suggest	this term is used in questions where students need to apply their knowledge and understanding to a new situation
Use	the answer must be based on the information given in the question. Unless the information given in the question is used, no marks can be given. In some cases you might be asked to use your own knowledge and understanding
Write	only a short answer is required, not an explanation or a description

1 Energy

Changes in Energy

This topic first covers some underlying principles of **energy** changes before going on to cover the use of a number of energy transfer equations and practical applications of energy changes.

Example

Which statement below is **incorrect**? Tick **one** box. *(1 mark)*

Energy can be transferred. ☐

Energy can be created. ☐

Energy can be stored. ☐

> There will be some multiple choice questions in the exam. This is a knowledge recall question that tests knowledge of a vitally important concept in the topic of Energy (and one that is going to be useful in future questions in this section!).

Energy can be created. ✓

> As well as not being able to be created, energy cannot be destroyed either. This means all the energy in the **Universe** has existed since the start of time and will exist until the end of time; it is just transferred into different forms.

Example

Use answers from the box to complete the sentences below. Each word can be used once, more than once or not at all. *(2 marks)*

| electrical | system | gravitational | heat |

A _____ is an object or a group of objects. There are changes in the way energy is stored when a _____ changes. When an electric kettle boils, _____ energy is being converted into _____ and sound energy.

Energy

A **system** is an object or group of objects. There are changes in the way energy is stored when a **system** changes. When an electric kettle boils, **electrical** energy is being converted into **heat** and sound energy. ✓ ✓

One incorrect answer –1 mark

> There may be 'fill in the blank' questions like this in the exam and this one seems straightforward with four words and four gaps. However, it is important to consider that the instructions say that each word can be used once, more than once or not at all. In this example one of the words isn't used at all whilst another is used twice.

> When considering energy, a system is just a group of objects. Energy is stored in the objects that make up a system, so when the system changes the way the energy stored within the system also changes.

Example

A group of scientists carried out an investigation into slingshots. In Experiment 1 a ball of **mass** 15 g was fired from a slingshot. It reached a maximum **speed** of 20 m/s.

a) Write down the equation that relates **kinetic energy**, mass and speed. *(1 mark)*

> There are certain equations that you need to recall and some that will be given on the Physics Equations Sheet in the exam. This equation is an example of one which you need to recall. It is very important that you learn these equations thoroughly as there may be a mark for correctly writing down the equation and then marks for applying the equation to the question.

kinetic energy = $0.5 \times$ mass $\times$ (speed)^2 ✓

b) Calculate the maximum kinetic energy of the ball. *(2 marks)*

> Questions will often not just be a straightforward substitution into a formula. In this case the kinetic energy equation uses kg as a unit of mass, but the question gives the mass in grams. You therefore need to convert the mass into kg before substituting the values into the equation.

$15 \text{ g} = 0.015 \text{ kg}$

kinetic energy = $0.5 \times 0.015 \times (20)^2$

$= 0.5 \times 0.015 \times 400$ ✓

$= 3 \text{ J}$ ✓

Energy

i) In Experiment 2, the elastic of the same slingshot was pulled back 0.35 m. The **elastic potential energy** of the slingshot elastic was 2.45 J and the **limit of proportionality** was not exceeded.

Determine the spring constant of the slingshot elastic.

Use the correct equation from the Physics Equations Sheet. *(3 marks)*

> The formula for elastic potential energy will be given to you in the Physics Equations Sheet in the exam paper (see pages 121–122 of this book). You could be asked to change the subject of an equation in the exam and this question requires you to do so in order to find the spring constant. In all calculations it is very important to show your working. However, it is particularly important when changing the subject of an equation as there will usually be a mark for the correct rearrangement, the correct substitution and the correct answer. Examiners will award the full mark allocation if you just write the correct answer, but this is a risky strategy because a wrong answer will score no marks whilst an incorrect answer with some correct working may score some marks.

elastic potential energy = 0.5 × spring constant × (extension)^2
spring constant = elastic potential energy ÷ (extension^2 × 0.5) ✓
= 2.45 ÷ (0.35^2 × 0.5)
= 2.45 ÷ (0.1225 × 0.5) ✓
= 40 N/m ✓

ii) Was the **extension** of the slingshot greater in Experiment 1 or Experiment 2?

Explain your answer. *(2 marks)*

> The exam papers will include lots of questions that ask you to apply your physics knowledge to experimental examples. In this case you have to deduce the extension of the slingshot in Experiment 1. The key to this question is the fundamental idea that energy cannot be created. The ball in Experiment 1 had a maximum kinetic energy of 3 J, so the elastic potential energy of the slingshot elastic must have been 3 J or greater in order to transfer 3 J of energy to the ball. This is greater than the elastic potential energy in Experiment 2. As the same slingshot was used in both investigations, the spring constants must have been the same. Therefore the only way to give a greater elastic potential energy is to have a greater extension.

The extension must have been greater in Experiment 1 ✓ as this must have had a greater elastic potential energy than in Experiment 2 in order to transfer 3 J of energy to the ball. ✓

Energy

Example

A plane is flying at a constant altitude (height) of 800 m. It begins its descent to a landing area in a field, lands and then moves along the field for a short distance.

a) Explain how the **gravitational potential** energy of the plane changes in the above situation. *(3 marks)*

> The gravitational potential energy of the plane remains constant whilst its altitude remains constant. ✓ As it descends to the field, its gravitational potential energy decreases. ✓ When the plane touches down on the field, its gravitational potential energy is zero. ✓

Whilst this question seems relatively straightforward, it requires you to understand that once the plane lands the height of the plane is zero, the gravitational potential energy is also zero and will remain zero whilst the plane is on the ground.

b) Calculate the gravitational potential energy of the plane when it is flying at 800 m. The mass of the plane is 650 kg and the gravitational field strength is 9.8 N/kg.

Write down any equations you use and give your answer in kilojoules to three significant figures. *(3 marks)*

The equation for gravitational potential energy is another equation you need to recall. You will always be given the gravitational field strength in a question; on Earth it is 9.8 N/kg. In this example, you are asked to give your answer in kilojoules as it is such a large value. A thousand joules is a kilojoule. You should make sure you are confident in converting between different magnitudes of SI units.

> gravitational potential energy (GPE) = mass × gravitational field strength × height ✓
> GPE = 650 × 9.8 × 800 ✓
> = 5 096 000 J
> = 5100 kJ (3 s.f.) ✓

Energy

Example

The table below shows the **specific heat capacity** of different substances.

Substance	Specific heat capacity (J/kg/°C)
Copper	390
Steel	460
Water	4180
Ethanol	2440

a) Determine the change in thermal energy when 0.32 kg of ethanol is heated from 34°C to 53°C.

Use the correct equation from the Physics Equations Sheet. *(2 marks)*

> First determine the temperature change before substituting values into the formula.

change in temperature = 53 − 34 = 19°C ✓
change in thermal energy = 0.32 × 2440 × 19
= 14 835.2 J ✓

b) If the same increase in thermal energy occurred in the same mass of water at 34°C, what temperature would it reach? Explain the differences in the values. *(4 marks)*

> To answer this question, you need to rearrange the equation so that temperature change is the subject and then substitute the values from part a) and the specific heat capacity of water from the table. Then explain why the values are different.

change in thermal energy = mass × specific heat capacity × temperature change

temperature change = change in thermal energy ÷ (mass × specific heat capacity) ✓

= 14 835.2 ÷ (0.32 × 4180) ✓

= 14 835.2 ÷ 1.3376

temperature change = +11°C

new water temperature = 45°C ✓

The water temperature didn't reach as high a value as the ethanol temperature, as water has a higher specific heat capacity than ethanol so it requires a greater increase in energy to cause the same rise in temperature. ✓

Energy

Power and Conservation and Dissipation of Energy

Example

When energy transfers take place in a closed **system**, what happens to the total energy of the system? Tick **one** box. *(1 mark)*

There is a net increase in total energy. ☐
There is a net decrease in total energy. ☐
There is no net change in total energy. ☐

> There is no net change in total energy. ✓

Net change means no overall change.

Example

A model crane was used in an investigation into **power**. The electrical motor on the crane was used to raise a **mass** of 0.5 kg to a height of 1.3 m in 7 seconds.

a) i) The **force** exerted by the crane is equal to the **weight** of the mass.
Calculate the force exerted by the crane.
Assume the gravitational field strength is 9.8 N/kg.
Write down any equations you use. *(2 marks)*

> weight = mass × gravitational field strength
> weight = 0.5 × 9.8 ✓
> weight = 4.9 N ✓

ii) Write down the equation that links **work** done, force and **distance** moved along the line of action of the force. *(1 mark)*

> work done = force × distance ✓

The units for this equation are work done in joules (J), force in newtons (N) and distance in metres (m).

Energy

iii) Calculate the work done by the model crane. *(2 marks)*

> work done = 4.9 × 1.3 ✓
> work done = 6.37 J ✓

Use the weight value from the answer for **i)** as the force value.

iv) Write down the equation that links power, work done and time. *(1 mark)*

> $power = \dfrac{work\ done}{time}$ ✓

The units for this equation are power in watts (W), work done in joules (J) and time in seconds (s).

v) Calculate the power of the model crane. *(2 marks)*

> $power = \dfrac{6.37}{7}$ ✓
> power = 0.91 W ✓

b) A second model crane raised the mass in 10 seconds. Compare the power of the first crane with the second crane.
Explain your answer. *(2 marks)*

> As the mass is staying the same but the time taken is increasing ✓ the second crane must have a lower power than the first crane ✓.

You could answer this question by calculating the power of the second model crane. There is however no need to do this as the crane is lifting the same mass in a longer time.

c) i) What useful energy transfer is occurring in this experiment? *(2 marks)*

> Electrical energy to kinetic energy ✓ to gravitational potential energy ✓.

As the question specifies a transfer you need to ensure you state at least **two** types of energy; the initial energy and the form it is **transferred** to. Refer back to the start of the question for the initial energy.

Energy

ii) Name **one** possible form of non-useful energy that may be released in this investigation. *(1 mark)*

Heat or sound ✓

> When asked to identify wasted energy first ensure you know what the useful energy transfer is! In a process involving moving parts there will be friction, which will release heat energy. Moving objects will also release sound energy.

d) **HT** Explain how a lubricant could be used to increase the **efficiency** of the cranes. *(2 marks)*

Applying lubricant to moving parts would decrease the friction between the moving parts ✓. This would decrease the energy lost as heat so make the transfer more efficient ✓.

> You are expected to be able to describe ways to increase the efficiency of an energy transfer. In a transfer involving moving parts lubricants would be a common way to do this.

Example

During 2 hours of operation a computer has a total energy input of 3240 kJ. It has a useful energy output of 2750 kJ.

a) Write down the equation that links efficiency, useful output energy transfer and total input energy transfer. *(1 mark)*

$$\text{efficiency} = \frac{\text{useful output energy transfer}}{\text{total input energy transfer}}$$ ✓

b) Calculate the efficiency of the computer.
Give your answer as a percentage. *(2 marks)*

$\text{efficiency} = \frac{2750}{3240}$ ✓

$\text{efficiency} = 0.85$

$\text{efficiency} = 0.85 \times 100 = 85\%$ ✓

> To convert the decimal to a percentage multiply it by 100.

Energy

c) An older version of the computer had the same total input energy transfer but an efficiency of 78%.
What was the useful output energy? (2 marks)

> useful output energy transfer = efficiency × total input energy transfer
> useful output energy transfer = 0.78 × 3240 ✓
> useful output energy transfer = 2527 kJ ✓

As the efficiency of the older computer is lower than that of the newer computer and its total input energy is the same, then the useful output energy must be lower than 2750 kJ.

Example

An investigation was carried out into the effectiveness of different insulation materials. The three different insulation materials were wrapped around three different containers of water. 100°C water was added to each container and then the temperature of the water was recorded at regular time intervals. The table below shows the results of the investigation.

Material	Temperature (°C)			
	Start	1 minute	2 minutes	3 minutes
A	100	95	87	80
B	100	99	96	95
C	100	99	97	93

a) Calculate the average change in temperature for each material.
Give your answer in °C per minute to two significant figures. (3 marks)

$A = \frac{(80 - 100)}{3}$
$A = \frac{-20}{3}$
$A = -6.7°C$ / minute ✓

$B = \frac{(95 - 100)}{3}$
$B = \frac{-5}{3}$
$B = -1.7°C$ / minute ✓

$C = \frac{(93 - 100)}{3}$
$C = \frac{-7}{3}$
$C = -2.3°C$ / minute ✓

> To calculate the average change in temperature, take the final temperature away from the start temperature and divide by the time taken.

Energy

b) Which of the three materials had the highest thermal **conductivity**? Explain your answer using the results from **a)**. *(2 marks)*

> Material A ✓, as this showed the largest average decrease in temperature per minute ✓.

c) Why was it important that the water in all three containers had the same initial temperature? *(1 mark)*

> To ensure the temperature changes can be compared with the same start point ✓.

> In questions like this you should avoid writing 'to make it a fair test'.

d) The experiment was repeated with exactly the same conditions but in a colder room. How would the results differ? *(2 marks)*

> The temperatures would decrease to lower levels ✓ as there is a greater difference between the room temperature and the initial temperature of the water. This would lead to a greater average temperature change per minute ✓.

Example

a) Explain why a homeowner would want to reduce the thermal conductivity of the walls of their home. *(2 marks)*

> Reducing thermal conductivity reduces heat loss from the house ✓, which will mean less energy is required to keep the house warm in the winter, which will save money on energy bills ✓.

b) A homeowner's heating system has a useful power output per day of 1.8 kW. The total power input per day is 2.1 kW.

  i) Write down the equation that links efficiency, useful power output and total power input. *(1 mark)*

  > $$\text{efficiency} = \frac{\text{useful power output}}{\text{total power input}} \checkmark$$

  ii) Calculate the efficiency of the heating system. Give your answer as a percentage. *(2 marks)*

  > $\text{efficiency} = \frac{1.8}{2.1}$ ✓
  > $\text{efficiency} = 0.86 \times 100 = 86\%$ ✓

iii) A new boiler increases the efficiency of the heating system to 91% whilst maintaining the same total power input.
What is the new useful power output per day? *(2 marks)*

> useful output energy transfer = efficiency × total input energy transfer
> useful output energy transfer = 0.91 × 2.1 ✓
> useful output energy transfer = 1.91 kW per day ✓

Convert the percentage efficiency into a decimal before substituting into the rearranged equation.

National and Global Energy Resources

This topic focuses on energy resources, comparing **renewable** and non-renewable energy resources, including their environmental impact, changing trends in their use and their suitability in different situations.

Example

a) Complete the table using the words in the box below. *(2 marks)*

| wind | wave | oil | tidal | natural gas |

Renewable	Non-renewable

Renewable	Non-renewable
wind	oil
wave	natural gas
tidal	
✓	✓

One mark for each correct column.

Renewable energy resources are those that can be replenished as they are used. All fossil fuels are examples of non-renewable energy resources.

Energy

b) A hospital is considering installing solar panels to provide electricity from a renewable source.

   i) Give **one** advantage of solar panels over generating electricity using fossil fuels. *(1 mark)*

   > Solar does not produce greenhouse gases such as carbon dioxide, whilst burning fossil fuels does ✓.

   > Solar is mentioned in the question as a renewable energy resource so you cannot use this as an advantage. You should also avoid subjective, simplistic statements such as 'cheap'. Instead, consider what is the big advantage of solar energy over fossil fuels.

   ii) Explain why it would be dangerous for the hospital to get all of its electricity from solar panels. *(2 marks)*

   > Solar panels do not provide a constant source of electricity ✓, which is required for the safe running of the hospital ✓.

   > In the exam you may be asked to comment on the appropriateness of different energy resources for different uses. In this case the issue is the reliability of solar panels.

Example

The table below shows how the UK's total energy consumption is divided between different sectors.

Sector	Energy consumption (%)
transport	40
industry	17
services	14
domestic	

a) Calculate the energy consumption of the domestic sector. *(1 mark)*

> domestic = 100 − 40 − 17 − 14 = 29% ✓

> As this data refers to total energy consumption, the percentages must add up to 100.

Energy

The domestic sector mainly uses natural gas and electricity generated from both renewable and non-renewable energy resources.

b) i) The UK consumed 90.73 terajoules (TJ) of energy during the year. What is this in joules? Give your answer in standard form to two significant figures. *(1 mark)*

> The prefix 'tera' denotes multiplying by 10^{12}.

90.73×10^{12} J $= 9.1 \times 10^{13}$ J ✓

ii) Determine the energy used by services during the year. Give your answer in terajoules (TJ) to two significant figures. *(2 marks)*

> To answer this question, multiply the total UK energy by the percentage used by the services sector.

Energy used by services sector $= 90.73 \times 17\%$

$= 90.73 \times 0.17$ ✓

$= 15$ TJ ✓

c) Explain how the energy resources used in transport may differ from this. *(1 mark)*

Transport mainly uses fuels derived from oil and a small amount of electricity ✓.

> You need to be able to compare the way different energy resources are used in electricity generation, transport and heating.

d) How do you predict that the energy resources used in transport and the domestic sector will change over the next 50 years? *(2 marks)*

In the domestic sector there will probably be an increase in the amount of electricity from renewable energy resources ✓. *In transport there will probably be a decrease in the use of fuels derived from oil and an increase in the use of electricity* ✓.

> You could be asked to consider trends and patterns in energy use; think particularly about how the balance between non-renewable and renewable energy resources has changed and will have to change in the future.

Energy

Example

Large batteries are being built to efficiently store electricity generated by wind farms. Explain how this development is helping to overcome some of the problems of electricity generation by wind farms. *(4 marks)*

Wind farms only generate electricity when the wind is blowing ✓. This makes them an unreliable energy resource as they cannot supply electricity consistently ✓. Large batteries could be used to store excess electricity generated and then release it into the grid when the wind is not blowing ✓. This would mean homes and businesses could still receive a consistent supply of electricity even when the wind is not blowing ✓.

This is an example of a 4-mark extended response question. Like the 6-mark extended response questions, a 4-mark extended response question is marked in levels. The table below shows the different level descriptors. In order for your answer to be placed in a level you have to satisfy the criteria laid out for that particular level. The quality of your answer then determines what mark you are awarded within that level. The key aspect of all these questions is developing a sustained line of reasoning that is coherent, relevant, supported by evidence (substantiated) and logically structured. This means you should focus on writing well structured answers with a logical order that relate directly to the question and do not contain any non-relevant material.

Level	Marks
Level 2: Full description of how the battery overcomes the most significant disadvantage of wind power and the importance of this in producing a consistent supply to the electricity grid.	3–4
Level 1: Basic description of the advantages of a battery with little explanation of how this would benefit consistent supply to the grid.	1–2
No relevant content.	0

For more on the topics covered in this chapter, see pages 26–29 of the *Collins GCSE AQA Physics Revision Guide*.

2 Electricity

Standard Circuit Diagram Symbols

This is a short topic that covers the different circuit symbols used to construct circuit diagrams.

Example

Draw the symbols of the following components. (3 marks)

a) Lamp
b) Variable resistor
c) Diode

a)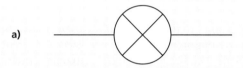

b)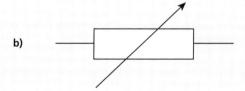

c)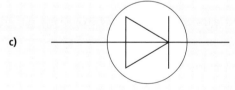

✓ ✓ ✓ One mark for each correct drawing.

You need to be able to identify, draw and interpret 14 different circuit symbols.

Electricity

Electrical Charge and Current

This is a short, fairly straightforward topic but it does contain some very important equations.

Example

a) Use the answers in the box below to complete the sentence. Each word can be used once, more than once or not at all. *(2 marks)*

> potential   charge   current   closed   resistance

For electrical charge to flow through a _____ circuit, the circuit must include a source of _____ difference. Electric current is a flow of electrical charge. The size of the electric _____ is the rate of flow of electrical _____.

> For electrical charge to flow through a **closed** circuit, the circuit must include a source of **potential** difference. Electric current is a flow of electrical charge. The size of the electric **current** is the rate of flow of electrical **charge**. ✓ ✓
>
> –1 mark for an incorrect answer.

In this question five options are given for four answers – this means you'll not use at least one of the words given.

b) Write down the equation that links **charge**, current and time. *(1 mark)*

> charge flow = current × time ✓

The units for this equation are charge flow in coulombs, current in amperes and time in seconds.

c) i) Calculate the charge flow at a point in the circuit where the current is 6 amps for 13 seconds. *(2 marks)*

> charge flow = 6 × 13 ✓
> charge flow = 78 coulombs (C) ✓

Electricity

ii) For the charge flow to change to 150 C, how long would the current have to be flowing? *(2 marks)*

$$time = \frac{charge\ flow}{current}$$

$$time = \frac{150}{6} \checkmark$$

$$time = 25\ s \checkmark$$

> The charge flow is greater than the answer to c) i) so the time must be greater than 13 seconds.

iii) The circuit is a single closed loop. What is the current at a second point on the circuit? *(1 mark)*

6 A ✓

> This question requires you to know that the current is the same in every point of a single closed loop (a series circuit).

Example

During a school physics experiment, a circuit is set up using a battery. The circuit is a single closed loop and an ammeter in the circuit shows a reading of 3 A. The total **resistance** of the circuit is 3 ohms.

a) Write down the equation that links **potential difference**, current and resistance. *(1 mark)*

$$potential\ difference = current \times resistance \checkmark$$

> The units of this equation are potential difference in volts, current in amperes and resistance in ohms. In the exam, questions will use the term potential difference. You will gain credit for the correct use of either potential difference or voltage but it's good practice to consistently use the term potential difference.

b) Calculate the resistance of the circuit. *(2 marks)*

$$potential\ difference = 3 \times 3 \checkmark$$

$$potential\ difference = 9\ V \checkmark$$

Electricity

c) A second battery, with the same potential difference as the original, is added to the circuit. The resistance in the circuit remains the same.
What is the new current flowing through the circuit? *(2 marks)*

$$current = \frac{potential\ difference}{resistance}$$

$current = \frac{18}{3}$ ✓

$current = 6A$ ✓

> As the potential difference has increased to 18 V whilst the resistance has remained constant, the current in the circuit will increase.

Resistors

In this topic you need to learn the shapes of the current / potential difference graphs of three resistors and be able to explain these shapes. You also need to explain how LDRs and thermistors can be used in practical applications.

Example

a) Identify the component that would produce the current / potential difference graph below and explain its shape. *(3 marks)*

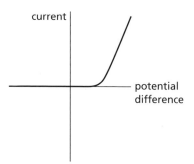

Diode ✓

The current through a diode flows in one direction only ✓ as a diode has a very high resistance in the reverse direction, as shown by the right side of the graph ✓.

> This is one of three potential difference current graphs you need to learn. They each have a characteristic shape due to the nature of the component. In each case you need to explain why the component produces the graph that it does.

Electricity

b) Explain why it would be wrong to describe this component as an **ohmic conductor**. *(2 marks)*

> The current through an ohmic conductor is directly **proportional** to the potential difference across the resistor ✓. This graph does not show a directly proportional relationship ✓.

> You need to learn the definition of an ohmic conductor.
> On a line graph a directly proportional relationship would be a straight line passing through the origin.

c) Explain how an LDR could be used to ensure outside lights only turn on during the hours of darkness. *(3 marks)*

> The resistance of an LDR increases as light intensity decreases ✓. The LDR should therefore be wired in a circuit that turns a light on if the current flowing through the circuit is low, as this will occur when the resistance of the LDR is high, which is in the dark ✓, and will not occur in the light, when the LDR resistance is low ✓.

Example

The table below shows the results into an investigation into the effect of temperature on the resistance in a circuit that contains a thermistor.

Temperature (°C)	Current (A)	Voltage (V)	Resistance (Ω)
20	3	8	2.7
30	6	8	Z
40	9	8	0.9

a) Calculate the missing resistance value, Z.
Show your working, including the equation used. *(3 marks)*

$$\text{resistance} = \frac{\text{potential difference}}{\text{current}} \checkmark$$

$$\text{resistance} = \frac{8}{6} \checkmark$$

$$\text{resistance} = 1.3\,\Omega \checkmark$$

> This question requires you to rearrange the following equation:
> potential difference = current × resistance

Electricity

b) Explain the pattern shown by the resistance values. (1 mark)

> As the temperature increases the resistance decreases ✓.

> When describing a trend, state the change in the independent variable (in this case temperature) and the corresponding change in the dependent variable (in this case resistance).

c) Describe how a thermistor like this one could be used in the thermostat of a heating system. (3 marks)

> As the temperature of the thermistor increases the resistance decreases ✓. The thermistor should be wired into a circuit that turns the heating system on when the current through the circuit is low ✓, as this will occur when the resistance of the thermistor is high in lower temperatures ✓.

Series and Parallel Circuits

This topic covers the differing properties of components wired in **parallel** and wired in **series**.

Example

The below circuit was set up during the testing of a new fuse design.

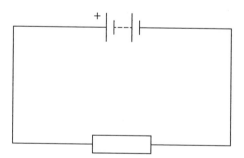

a) Identify this type of circuit. Explain your answer. (2 marks)

> A series circuit ✓, as the components are wired in a single closed loop ✓.

> Electrical components can be joined in series or in parallel. Some circuits will be either series or parallel but some circuits will have both series and parallel parts.

Electricity

) Explain what components would need to be added to the circuit in order to find the resistance of the fuse. **(2 marks)**

> Wire an ammeter in series with the fuse to find the current drawn by it ✓ and a voltmeter in parallel to find the potential difference across the fuse ✓.

In order to calculate resistance of a component you need to find the current drawn by the component and the potential difference across it.

xample

he circuit below was built during a physics practical in a school lab.

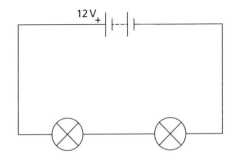

) The resistance of each of the lamps is 3 Ω. What would be the total resistance of the circuit?
Write down any equations you use. **(2 marks)**

> $R_{total} = R_1 + R_2$ ✓
> $R_{total} = 3 + 3$
> $R_{total} = 6\ \Omega$ ✓

In a series circuit the total resistance of two components is the sum of the resistance of each component.

Electricity

b) The circuit is modified as shown below. Each of the meters has a negligible resistance

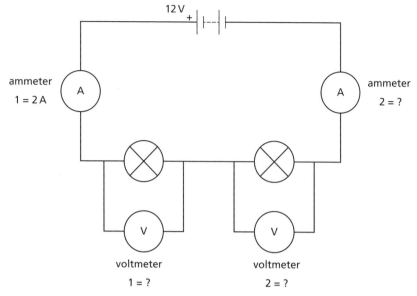

Calculate the readings that would be shown on the following meters. In each case explain how you arrived at your answer.

i) Ammeter 2 (2 marks)

2A ✓, as it would be the same reading as ammeter 1 ✓.

ii) Voltmeter 1 and voltmeter 2 (2 marks)

Both would have a reading of 6V ✓, as the potential difference of the supply would be shared between each of the lamps ✓.

> In a series circuit the current is the same through each component and the total potential difference of the power supply is shared between the components. As the meters all have a negligible resistance you do not need to consider them in your answer.

Electricity

Example

The circuit below was constructed during an investigation into the resistance of different components.

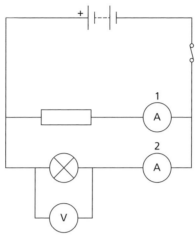

a) The voltmeter gives a reading of 9 V. What would the potential difference be across the resistor?
   Explain how you arrived at your answer. *(2 marks)*

   The potential difference across the resistor would be 9V ✓, as it would be the same as the potential difference across the lamp, as this is a parallel circuit ✓.

   In a parallel circuit the potential difference across each component is the same.

b) Ammeter 1 gives a reading of 3 A and ammeter 2 gives a reading of 6 A. What would be the total current through the whole circuit?
   Explain how you arrived at your answer. *(2 marks)*

   The total current would be 9A ✓, as the total current in a parallel circuit is the sum of the currents through the branches ✓.

   The total current in a parallel circuit is the sum of the currents through the separate branches.

Electricity

c) Predict the effect of adding additional resistors in parallel to this circuit. How would this effect be different if this was a series circuit? In both cases explain your answer. *(4 marks)*

> Adding additional resistors in parallel would decrease the total resistance of this circuit ✓, as the total resistance of a parallel circuit is less than the resistance of the smallest individual resistor ✓. In a series circuit adding additional resistors increases the total resistance ✓, as the total resistance is the sum of the resistance of each individual resistor ✓.

Domestic Uses and Safety

For all areas of this topic it's important to explain how certain features improve safety and reduce risk.

Example

a) The graph below shows the potential difference in a circuit.

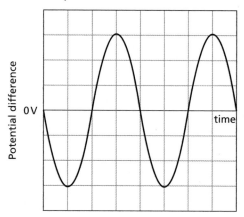

Is this an example of **a.c.** or **d.c.**?
Explain how you arrived at your answer. *(2 marks)*

> a.c. ✓, as the potential difference is alternating ✓.

Electricity

b) Use the answers in the box below to complete the sentences. Each possible answer can be used once, more than once or not at all. *(2 marks)*

| a.c. | Hz | 230 | 50 | d.c. | 60 |

Mains electricity is an _____ supply. In the United Kingdom the domestic electricity supply has a **frequency** of _____ and is about _____ V.

Mains electricity is an **a.c.** supply. In the United Kingdom the domestic electricity supply has a frequency of **50 Hz** and is about **230** V. ✓ ✓
One incorrect answer −1 mark.

c) i) Complete the table of the insulation colours of the wires in the three-core cables that connect most electrical appliances to the mains. *(3 marks)*

Insulation colour	Wire
brown	
blue	
green and yellow stripes	

Insulation colour	Wire	
brown	live	✓
blue	neutral	✓
green and yellow stripes	earth	✓

One incorrect answer −1 mark.

ii) Explain why it's important that the insulation of wires is colour coded. *(2 marks)*

So they can be easily identified ✓, allowing a plug to be wired correctly. It also allows easy identification of the potentially dangerous live wire ✓.

iii) Describe the differences between the potentials of the live and the neutral wires. Explain your answer. *(2 marks)*

The potential difference between the live wire and the neutral wire is about 230 V ✓ as the neutral wire is at earth potential 0 V ✓.

Electricity

   iv) In what situation would the earth wire carry a current?   (1 mark)

   If there was a fault with the appliance ✓

Energy Transfers

This is a wide ranging topic covering power, energy transfers in everyday appliances and the **National Grid**.

Example

A games console draws a current of 5 A and a resistance of 10 ohms.

a) Write down the equation that relates power, current and resistance.   (1 mark)

> *power = (current)² × resistance* ✓
>
> The units of this equation are power in watts, current in amperes and resistance in ohms.

b) Calculate the power of the games console.   (2 marks)

> *power = (current)² × resistance*
> *power = 5² × 10*
> *power = 25 × 10* ✓
> *power = 250 W* ✓

c) An update to the games console improves its efficiency. It now has a power of 230.4 W.
   Assuming its resistance has not changed, what current is it now drawing?   (3 marks)

> $current^2 = \frac{power}{resistance}$
> $current^2 = \frac{230.4}{10}$
> *current² = 23.04* ✓
> *current* = $\sqrt{23.04}$ ✓
> *current* = 4.8 A ✓
>
> As the power rating has decreased whilst the resistance has remained the same, the current must be lower than 5 A.

Electricity

d) A new games console draws a current of 15 A and has a potential difference of 10 V. Does this new console have a greater or smaller power than the updated console?
Write down any equation you use. *(4 marks)*

> power = potential difference × current ✓
> power = 10 × 15 ✓
> power = 150 W ✓
> This is smaller than the power of the updated console. ✓

> This question requires you recall and apply a different power equation to the one in **a)**.

e) An update of the new console changes the current it draws so it now has a power of 125 W.
Assuming the potential difference is unchanged, what is the new current drawn by this games console? *(2 marks)*

> $current = \dfrac{power}{potential\ difference}$
> $current = \dfrac{125}{10}$ ✓
> current = 12.5 A ✓

Example

a) Which of the below does the amount of energy and appliance transfers **not** depend on? Tick **one** box. *(1 mark)*

How long the appliance is switched on for. ☐
The power of the appliance. ☐
The age of the appliance. ☐

> The age of the appliance. ✓

b) Whilst cooking a dish a microwave has a charge flow of 1130 coulombs and a potential difference of 120 V.

   i) Write down the equation that links energy transferred, potential difference and charge flow. *(1 mark)*

35

Electricity

energy transferred = charge flow × potential difference ✓

> The units of this equation are energy transferred in joules, charge flow in coulombs and potential difference in volts.

ii) Calculate the energy transferred in the microwave.
Give your answer in kJ to three significant figures. *(2 marks)*

energy transferred = 1130 × 120
energy transferred = 135 600 J ✓
135 600 in kJ to 3 s.f. = 136 kJ ✓

iii) When cooking the same dish, a newer model of this microwave transfers 110 kJ while the potential difference remains unchanged.
What is the charge flow of the new model?
Give your answer to three significant figures. *(2 marks)*

$$\text{charge flow} = \frac{\text{energy transferred}}{\text{potential difference}}$$
$$\text{charge flow} = \frac{110\,000}{120} \checkmark$$
charge flow = 917 C ✓

> As the energy transferred has decreased while the potential difference has remained the same, the charge flow must be below 1130 C. Convert the energy transferred into joules before substituting the value into the rearranged equation.

iv) The newer model of the microwave has a power of 1530 W. How long was the dish cooked for?
Give your answer in whole seconds and show how you arrived at your answer, including writing down any equations you use. *(3 marks)*

$$\text{time} = \frac{\text{energy transferred}}{\text{power}} \checkmark$$
$$\text{time} = \frac{110\,000}{1530} \checkmark$$
time = 72 s ✓

> To answer this question you need to rearrange the energy transfer equation: energy transferred = power × time. Ensure you convert the energy transferred from kJ to J before substituting the value into the equation.

Electricity

c) The table below shows the power ratings of different electrical appliances.

Appliance	Power rating
coffee maker	1400
dishwasher	1500
food blender	400
fridge / freezer	400
washing machine	500

Which of the appliances brings about the greatest transfer of energy in 10 minutes of use?
Explain your answer. *(2 marks)*

The dishwasher ✓, as it has the greatest power rating so would transfer the greatest amount of energy in the given time ✓.

Example

a) Which of the following statements is true? Tick **one** box. *(1 mark)*

Electrical power is transferred from power stations to consumers using the National Grid. ☐

Electrical power is transferred from consumers to power stations using the National Grid. ☐

Electrical force is transferred from power stations to consumers using the National Grid. ☐

Electrical power is transferred from power stations to consumers using the National Grid. ✓

b) Describe the use of step-up and step-down **transformers** in the National Grid. *(2 marks)*

Step-up transformers increase the potential difference from the power station to the transmission cables ✓. Step-down transformers are used to decrease the potential difference for use in homes ✓.

Electricity

c) Explain why transformers in the National Grid are important for:

i) efficient transfer of energy *(2 marks)*

> Step-up transformers increase potential difference for transmission by power cables as they reduce the current ✓, so reducing heat loss from the cables ✓.

ii) safe use of electricity. *(1 mark)*

> Step-down transformers decrease the potential difference to a much lower level that is safe to use in appliances in the home ✓.

Static Electricity

This is a small topic that covers static charge and electrical fields. The concepts in this topic explain phenomena such as static electricity and **sparking**.

Example

In an investigation into static charge a cloth is rubbed vigorously over a perspex rod. The perspex rod is then observed to become attracted to an object with a negative charge.

a) Explain this observation. *(3 marks)*

> When the rod is rubbed with the cloth electrons are transferred from the rod to the cloth ✓. As electrons are negatively charged this means the rod becomes positively charged ✓. A positively charged object is attracted to a negatively charged object ✓.

When certain insulating materials are rubbed against each other they become electrically charged due to negatively charged **electrons** that are rubbed off one material on to the other.

b) The force of **attraction** seen in this investigation is an example of **non-contact force**.
Explain what is meant by the term non-contact force. *(1 mark)*

> Non-contact forces do not require objects to be physically touching ✓.

There are a number of examples of different contact and non-contact forces you need to learn. These are covered in detail in the Forces section of the specification.

Electricity

c) The diagram shows three charged objects. Objects Y and Z have the same charges.

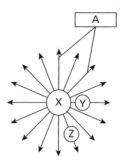

i) What is represented by the arrows labelled A? *(1 mark)*

   The object's electrical field ✓.

ii) Is object X positively or negatively charged?
    Explain your answer. *(2 marks)*

   It is positively charged ✓ because the arrows are pointing away from the object ✓.

> The type of charge on an object can be determined by the direction of the field lines of its electrical field. The direction of the field lines shows the direction a positive charge would move in the electrical field.

iii) Which of the objects would experience the greatest force?
     Explain how you arrived at your answer. *(2 marks)*

   Object Y ✓, as it is closer to object X than object Z. The force gets stronger as the distance between objects decreases ✓.

Example

When a plane is being fuelled with highly flammable fuel, earthing lines are used to prevent the build-up of static charge. Use the concept of electrical fields to explain the importance of these earthing lines. *(3 marks)*

If static charge builds up on the plane and fuelling equipment ✓, the interaction between the electrical fields could lead to sparking ✓. The sparking could ignite the flammable fuel ✓.

> Two charged objects with interacting electric fields can lead to sparking.

For more on the topics covered in this chapter, see pages 54–65 of the *Collins GCSE AQA Physics Revision Guide*.

3 Particle Model of Matter

Changes of State and the Particle Model

This short topic covers density and changes of state. You need to be able to recall and apply the pressure equation, explain the differences in density between the different **states of matter** in terms of the arrangement of **atom**s or molecules, and appreciate the differences between physical changes and chemical changes.

Example

a) Match the terms below with the correct **particle** model diagrams. *(2 marks)*

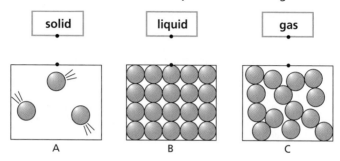

A: *gas*
B: *solid*
C: *liquid*
✓ ✓ One incorrect answer −1 mark.

b) Which of the substances pictured will have the lowest **density**? Explain how you arrived at your answer. *(2 marks)*

The gas (A) ✓, *as the particles are the furthest apart* ✓.

Example

a) Write down the equation that links density, mass and volume. *(1 mark)*

$$density = \frac{mass}{volume}$$ ✓

The units of this equation are density in kg/m³, mass in kg and volume in m³.

Particle Model of Matter

a) Calculate the density of a rock sample with a mass of 6.7 kg and a volume of 0.001 m³. *(2 marks)*

$density = \frac{6.7}{0.001}$ ✓

$density = 6700 \, kg/m^3$ ✓

b) Another, more porous rock sample with the same volume has a density of 5800 kg/m³.
What is the mass of this rock? *(2 marks)*

$mass = density \times volume$

$mass = 5800 \times 0.001$ ✓

$mass = 5.8 \, kg$ ✓

> As this rock has a smaller density in the same volume than the rock in **b)** then the mass value must be lower than 6.7 kg.

Example

a) Which of the following statements is correct? Tick **one** box. *(1 mark)*

Changes of state are a type of chemical change. ☐
Changes of state are physical changes. ☐
A material cannot recover its original properties after a physical change occurs. ☐

Changes of state are physical changes. ✓

b) A 7 kg sample of water freezes completely to form ice.
What is the mass of the ice sample?
Explain how you arrived at your answer. *(2 marks)*

The ice has a mass of 7 kg ✓; *this is because in a change of state the mass is conserved.* ✓

> Water freezing is an example of a change of state that is a physical change as opposed to a chemical change.

Particle Model of Matter

Internal Energy and Energy Transfers

The most important elements of this topic relate to changes in **internal energy**, particularly during changes of state. Make sure you don't mix up specific heat capacity and specific latent heat!

Specific heat capacity is the amount of energy required to raise the temperature of 1 kilogram of a substance by 1 degree Celsius. Specific latent heat is the amount of energy required to change the state of 1 kilogram of a substance with no change in temperature.

Example

An investigation was carried out into a solid substance that sublimes when heated. The solid was heated consistently for 10 minutes.

a) What **two** types of energy make up the internal energy of this system? *(2 marks)*

> Kinetic energy ✓ and potential energy ✓.
>
> Internal energy is the energy stored inside a system by the particles that make up the system.

b) What happens to the internal energy of the system during this investigation? Explain your answer. *(1 mark)*

> As it is being heated the internal energy increases ✓.

The graph below shows the results of the investigation.

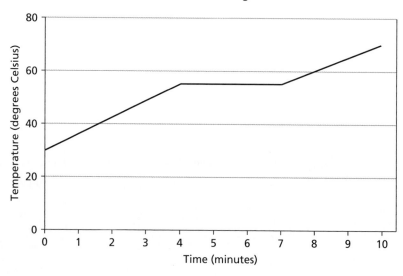

Particle Model of Matter

) Explain the shape of the graph between 4 and 7 minutes. *(2 marks)*

> The temperature is constant at this point even though the substance is being heated ✓ because the substance is changing state from a solid to a gas ✓.

When a substance sublimes it changes state from a solid to a gas.

Example

a) What is the difference between the specific **latent heat of fusion** and the specific **latent heat of vaporisation**? *(2 marks)*

> The specific latent heat of fusion is the energy required to change 1 kg of a substance from a solid to a liquid ✓. The specific latent heat of vaporisation is the energy required to change 1 kg of a substance from a liquid to a vapour ✓.

The specific latent heat of a substance is the amount of energy required to change the state of 1 kg of the substance, with no change in temperature.

b) The specific latent heat of helium is 0.021 J/kg.
What energy would be required to change 780 g from a liquid to a vapour?
Use the correct equation from the Physics Equations Sheet.
Give your answer to two significant figures. *(2 marks)*

> energy for a change of state = mass × specific latent heat
> energy for a change of state = 0.78 × 0.021 ✓
> energy for a change of state = 0.016 J ✓

c) 0.35 J of energy is required to change the same mass of hydrogen from liquid to a vapour.
What is the specific latent heat of hydrogen?
Give your answer to two significant figures. *(2 marks)*

> specific latent heat = $\frac{\text{energy for a change of state}}{\text{mass}}$
> specific latent heat = $\frac{0.35}{0.78}$ ✓
> specific latent heat = 0.45 J/kg ✓

As the energy value is greater than the answer for **b)**, this shows that the specific latent heat of hydrogen must be higher than that of helium.

Particle Model of Matter

Particle Model and Pressure

This is a short topic covering the behaviour of gases but it does contain an equation that is quite challenging to apply.

Example

An investigation into the properties of a gas was carried out. A sample of gas was sealed into a gas syringe.

a) The gas was heated, causing a rise in temperature from 45°C to 55°C. What effect would this have on the average kinetic energy of the molecules of the gas? (1 mark)

> The average kinetic energy of the molecules will increase ✓.

b) The gas remained at a volume of 130 cm³ while it was heated. What would happen to the **pressure** of the gas while it was heated? (1 mark)

> The pressure would increase ✓.

c) After heating, the gas was allowed to cool to a constant room temperature. The pressure of the gas was 200 kPa. The plunger of the gas syringe was unsealed and pulled outwards, increasing the volume of gas from 0.0003 m³ to 0.0004 m³.

　i) What would happen to the pressure of the gas? (1 mark)

> The pressure would decrease ✓.

　ii) Calculate the pressure of gas after the plunger was pulled out. (3 marks)

> $P_1 \times V_1 = P_2 \times V_2$
> $200 \times 0.0003 = P_2 \times 0.0004 \, m^3$ ✓
> $0.06 = P_2 \times 0.0004 \, m^3$
> $P_2 = \frac{0.06}{0.0004}$ ✓
> $P_2 = 150 \, kPa$ ✓

> As the units of pressure given in the question are kPa you should give your answer in kPa.

Particle Model of Matter

iii) In a follow-up investigation (where the temperature remained constant) the pressure of the gas changed from 180 kPa to 300 kPa. The initial volume of the gas was the same as in the first experiment.
What was the final volume of the gas? (3 marks)

$$P_1 \times V_1 = P_2 \times V_2$$
$$180 \times 0.0003 = 300 \times V_2 \checkmark$$
$$0.054 = 300 \times V_2$$
$$V_2 = \frac{0.054}{300} \checkmark$$
$$V_2 = 0.00018 \, m^3 \checkmark$$

d) **HT** Pushing the gas syringe plunger in and out repeatedly led to a small temperature change in the gas.
Explain why. (3 marks)

Pushing the plunger in and out repeatedly was transferring force ✓ to, so therefore doing work on, the gas ✓. This led to an increase in the internal energy of the gas and therefore an increase in the temperature of the gas ✓.

For more on the topics covered in this chapter, see pages 84–85 of the *Collins GCSE AQA Physics Revision Guide*.

4 Atomic Structure

Atoms and Isotopes

A key part of this topic is how the model of the atom has developed over time. You should be able to explain how new experimental evidence may lead to a scientific model being changed or replaced.

Example

a) Which of the below is the correct, approximate radius of an atom?
   Tick **one** box. *(1 mark)*

   1×10^{-2} metres ☐

   1×10^{-20} metres ☐

   1×10^{-10} metres ☐

> 1×10^{-10} *metres* ✓
>
> The radius of a **nucleus** is less than $\frac{1}{10000}$ of the radius of an atom. Most of the mass of an atom is concentrated in the nucleus.

b) Use the answers in the box to complete the passage below. Each word can be used once, more than once or not at all. *(2 marks)*

   | positively | negatively | atom | nucleus | electrons | protons |

   The basic structure of an _____ is a _____ charged _____ composed of both _____ and **neutrons** surrounded by _____ charged _____.

> The basic structure of an **atom** is a **positively** charged **nucleus** composed of both **protons** and neutrons surrounded by **negatively** charged **electrons**. ✓✓
>
> One incorrect answer –1 mark.

Atomic Structure

) The diagram below shows the structure of an atom.

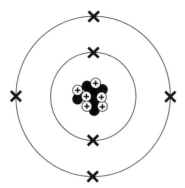

i) Explain how this structure would alter if the atom **absorbed** electromagnetic radiation. *(2 marks)*

> As electrons are absorbing electromagnetic radiation they would gain energy ✓ and jump to higher energy levels further from the nucleus ✓.

The electron arrangements of an atom may change with the absorption or the emission of electromagnetic radiation.

ii) All atoms of this **element** would have the same number of one of the particles found in the nucleus.
Identify this particle. *(1 mark)*

> All atoms of this element would have the same number of protons ✓.

The number of protons in an atom is called the **atomic number**.

iii) This element has an atomic number of 6 and a total of 12 particles in its nucleus.
How many neutrons are there in the nucleus? *(1 mark)*

> There are 6 neutrons in the nucleus ✓.

The number of protons + the number of neutrons = the mass number of the atom.

Atomic Structure

Example

Below are the three naturally occurring **isotopes** of hydrogen.

$$^{1}H \quad ^{2}H \quad ^{3}H$$
protium deuterium tritium

a) Explain the differences in the mass numbers. *(2 marks)*

These isotopes have the same number of protons (one) and different numbers of neutrons ✓. Protium has zero neutrons, deuterium has one neutron, and tritium has two neutrons ✓.

> Atoms of the same element can have different numbers of neutrons, these atoms are called isotopes of that element.

b) Why are all of these atoms known as hydrogen when they have different mass numbers? *(1 mark)*

They all have the same number of protons in the nucleus, this makes them the same element ✓.

c) None of these isotopes are **ions**. Use this information to determine the number of electrons in each isotope.
Explain how you arrived at your answer. *(2 marks)*

Each isotope has one electron ✓, as they are not ions they are uncharged so must have the same number of protons as electrons ✓.

Example

Before the discovery of the electron, atoms were thought to be tiny spheres that could not be divided.

a) Explain how the discovery of the electron led to this model being discarded and the plum pudding model being adopted. *(3 marks)*

The discovery of the electron meant that scientists realised that the atom contained smaller particles ✓. The plum pudding model suggested that the atom is a ball of positive charge ✓ with negative electrons embedded in it ✓.

Atomic Structure

The diagram below shows the results of the **alpha** particle scattering experiment carried out at Manchester University in 1909.

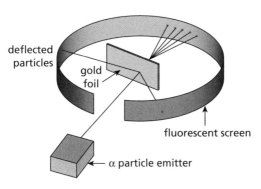

Rutherford, Geiger and Marsden's alpha scattering experiment

b) Using the diagram, explain how this experiment led to the rejection of the plum pudding model and the acceptance of a new model. *(4 marks)*

The alpha particle scattering experiment showed that most alpha particles were passing straight through the gold foil missing the nuclei of the atoms ✓, some were slightly deflected by passing close to the nuclei of the atoms ✓ and a small number were largely deflected due to striking the nucleus ✓. This led to the conclusion that the mass of an atom was concentrated in the centre at the nucleus and that the nucleus was charged. This is the nuclear model ✓.

There has been lots of further experimental work on the atom, including by Niels Bohr and James Chadwick.

c) Has the nuclear model been rejected due to this experimental work? Explain your answer, including examples. *(3 marks)*

No, the nuclear model has been modified ✓, e.g. Bohr's work led to the discovery that electrons orbit the nucleus at specific distances ✓. Chadwick's work provided the evidence to show the existence of neutrons within the nucleus ✓.

You are not required to know the details of Chadwick's experimental work or the experimental work supporting the Bohr model.

Atomic Structure

Atoms and Nuclear Radiation

This is a fairly wide-ranging topic covering a variety of different aspects of nuclear radiation. You should be able to balance nuclear equations, determine the **half-life** of a radioactive isotope from provided data, and explain the difference between radioactive contamination and irradiation.

Example

a) Explain the difference between **activity** and count-rate. *(2 marks)*

> Activity is the rate at which a source of **unstable** nuclei decays ✓.
> Count-rate is the number of decays recorded each second by a detector ✓.

A Geiger-Muller tube is an example of a detector that can be used to measure count-rate.

b) An investigation was carried out into emission by **radioactive** sources. The radioactive sources **emitted** either alpha particles, **beta** particles or **gamma** rays. A detector was used to detect radiation emitted by different sources at a range of distances in the air. The results of the investigation are shown in the table below. A tick represents radiation being detected at a particular distance.

Source	Distance / cm			
	2	4	6	8
A	✓	✓	✗	✗
B	✓	✓	✓	✓
C	✓	✓	✓	✓
D	✓	✓	✗	✗

i) What conclusions can be drawn from the results of this investigation? Explain your answer. *(3 marks)*

> A and D are both emitting alpha particles ✓ as they are absorbed by a few centimetres of air so are not detected after 6 cm ✓. B and C could be emitting beta particles or gamma rays ✓.

Atomic Structure

ii) Design a follow-up investigation to determine the radiation emitted by all of the sources. *(4 marks)*

> Set up a thin sheet of aluminium in front of sources B and C ✓. Place the detector on the other side of the aluminium sheet ✓. If radiation is detected on the other side of the aluminium sheet then the source must be emitting gamma rays ✓ as beta particles are absorbed by a thin sheet of aluminium ✓.

> This follow-up investigation will require a method of distinguishing between beta particles and gamma rays.

iii) What type of nuclear radiation emission is not being investigated in this experiment? *(1 mark)*

> Neutron emission ✓

iv) What are the main hazards associated with this investigation? *(1 mark)*

> Irradiation or contamination from the radioactive sources ✓.

v) Outline one method of reducing these hazards. *(1 mark)*

> Ensure suitable protective clothing and equipment is used when handling the sources ✓.

Example

a) Identify the types of radioactive decay that would emit the particles shown below. *(2 marks)*

A $_{-1}^{0}e$ B $_{2}^{4}He$

> A: beta decay ✓
> B: alpha decay ✓

> An alpha particle is a helium nucleus and a beta particle is an electron.

Atomic Structure

b) Explain the differing effects of alpha and beta decay on the mass and charge of the nucleus. **(4 marks)**

> Alpha decay causes both the mass and charge of the nucleus to decrease ✓ because the two protons and two neutrons are released ✓. Beta decay does not change the mass of the nucleus as a neutron is converted to a proton ✓ but it does cause the charge of the nucleus to increase ✓.

c) What type of emission does not cause the mass or the charge of the nucleus to change? **(1 mark)**

> Gamma rays ✓

d) Complete the equations below.
Identify each example as either alpha or beta decay. **(3 marks)**

A $\quad ^{\Box}_{38}Sr \longrightarrow ^{90}_{\Box}Y + ^{0}_{-1}e$

B $\quad ^{190}_{\Box}Pt \longrightarrow ^{\Box}_{76}Os + ^{\Box}_{2}He$

A $\quad ^{90}_{38}Sr \longrightarrow ^{90}_{39}Y + ^{0}_{-1}e \qquad$ *This is beta decay.*

B $\quad ^{190}_{78}Pt \longrightarrow ^{186}_{76}Os + ^{4}_{2}He \qquad$ *This is alpha decay.*

✓ ✓ ✓ −1 mark for each incorrect answer.

> You should be able to use the names and symbols of common nuclei and particles to write balanced equations that show single alpha (α) and beta (β) decay. You only need to be able to balance the atomic numbers and mass numbers, you won't need to identify the daughter elements (the elements that are produced by the decay).

Atomic Structure

Example

a) Use the answers in the box below to complete the following passage. Each answer can be used once, more than once, or not at all. *(2 marks)*

radiation	isotope	half-life	half	decay	gamma	nuclei

Radioactive _____ is random. The _____ of a radioactive isotope is the time it takes for the number of _____ of the isotope in a sample to halve, or the time it takes for the count-rate (or activity) from a sample containing the _____ to fall to _____ its initial level.

> Radioactive **decay** is random. The **half-life** of a radioactive isotope is the time it takes for the number of **nuclei** of the isotope in a sample to halve, or the time it takes for the count-rate (or activity) from a sample containing the **isotope** to fall to **half** its initial level. ✓ ✓
> One incorrect answer −1 mark.

b) The graph below shows the decay of a radioactive isotope.

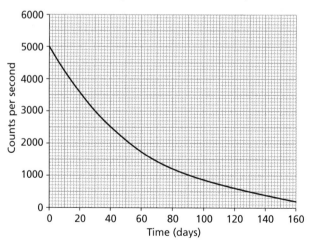

i) Estimate the half-life of this isotope.
   Explain how you arrived at your answer. *(2 marks)*

> The half-life of this isotope is 40 days ✓, as this is the time it took for the activity (counts per second) to halve ✓.

53

Atomic Structure

ii) Estimate the counts per second of the sample after 160 days. *(1 mark)*

counts per second after 80 days = $\frac{2500}{2}$ = 1250

counts per second after 120 days = $\frac{1250}{2}$ = 625

counts per second after 160 days = $\frac{625}{2}$ = 312.5 counts per second ✓

c) During the Chernobyl nuclear accident the town of Pripyat was covered in radioactive material ejected from the exposed core of the reactor of the nearby nuclear power plant. No one is able to live in Pripyat but people can visit and work there for short periods.

i) The Chernobyl disaster occurred in 1986. Explain why Pripyat is still dangerous. *(2 marks)*

*The area has become **contaminated** by the presence of radioactive atoms ✓, which are still decaying and emitting radiation ✓.*

ii) Explain why people can spend short amounts of time in Pripyat but cannot live there permanently. *(2 marks)*

People spending short periods of time at Pripyat are only receiving small doses of radiation ✓. People living there permanently would receive dangerously high doses of radiation over time ✓.

iii) A scientist visited Pripyat during a research trip. She was wearing a radiation detector on her coat. This showed her coat was absorbing nuclear radiation. When she returned to her lab she passed a radiation detector over the coat and the boots she was wearing. Her coat was not emitting radiation but her boots were.
Explain these observations. *(2 marks)*

*Her coat had been **irradiated** whilst in Pripyat so was exposed to nuclear radiation but has not become radioactive ✓. Her boots must have become contaminated with radioactive material ✓.*

iv) Why is it important that the scientist publishes any findings from her studies? *(1 mark)*

So her results can be shared with other scientists and checked by peer review ✓.

Atomic Structure

Hazards and Uses of Radioactive Emissions and of Background Radiation

This is a relatively short topic but there is a lot of scope for exam questions that ask you to apply your knowledge to specific real-world situations.

Example

Use the answers in the box to complete the table, to identify sources of background radiation as either natural or man-made. *(2 marks)*

| nuclear weapon testing | cosmic rays | rocks | nuclear accidents |

Natural	Man-made

Natural	Man-made	
rocks	nuclear accidents	✓
cosmic rays	nuclear weapon testing	✓

Example

The table below shows the mean dose levels of radiation in a year for different occupations.

Occupation	Mean dose (mSv)
doctor	5
underground miner	7.8
office worker	2.7

Atomic Structure

a) Explain these results. *(3 marks)*

> Miners have the highest mean dose, due to working underground surrounded by rocks which are natural sources of background radiation ✓. The doctor has the next highest dose as there are many sources of radiation in hospitals and doctors' surgeries ✓. The office worker has the lowest dose as their job does not bring them into contact with major sources of background radiation ✓.

> The level of background radiation and the radiation dose a person receives may be affected by their occupation and / or location.

b) The legal limit of radiation dose for a worker is 0.38 mSv a week. Should any of these workers be concerned about the dose they're receiving? Explain your answer. *(2 marks)*

> No, as 0.38 mSv per week is equivalent to 20 mSv a year ✓, which is much higher than all the doses these workers are receiving ✓.

> The dose given in the table is per year whilst the dose in the question is per week.

Example

Scientists were considering four different radioactive isotopes to use as a medical **tracer**.

Radioisotope	Half-life	Radiation emitted
1	2 years	α
2	3 months	γ
3	8 hours	γ
4	9 hours	α

Which isotope would be best suited for use as a tracer?
Explain how you arrived at your answer. *(2 marks)*

> Radioisotope 3 ✓, as it has a short half-life so can be easily detected and will not remain in the body for too long ✓ and it emits gamma radiation, which can pass through body tissues and be detected outside the body, unlike alpha radiation ✓.

> α is the symbol for alpha radiation and γ is the symbol for gamma radiation.

Atomic Structure

Example

Prostate cancer is the most common cancer in men in the UK – over 40 000 cases are diagnosed every year. Prostate cancer can be treated using external-beam radiation therapy. The tumour is targeted with a focused beam of radiation that destroys it. When used against prostate cancer, external-beam radiation therapy has a 95% cure rate.

The chance of mild to moderate damage to other tissues, leading to side effects, is approximately 0.0015 per radiation session. If side effects do develop the treatment is stopped.

Treatment of prostate cancer requires 40 radiation sessions.

Use this information to evaluate the benefits and risks of treating prostate cancer with external-beam radiation therapy. *(4 marks)*

> The benefits of the treatment are that it has a very high cure rate, with 95% of cases of prostate cancer being cured. The treatment does require a high number of radiation sessions but as the risk of serious side effects is very small (0.0015) per session, this does mean that the total risk of side effects for the whole treatment is still very small: $0.0015 \times 40 = 0.06$. This means the probability of side effects is only 6% for the whole treatment whilst there is a 95% chance of being cured. ✓✓✓✓

This is an example of a 4-mark extended response question. In this challenging question it is important you use all the information available to you in your answer. Whilst the total probability of being cured is given, the chance of side effects is only given per session. To determine the total chance of side effects, multiply the risk per session by the number of sessions. Multiply this number by 100 to give a percentage that can then be compared with the chance of being cured.

Level	Marks
Level 2: A clear comparison is made between the chance of being cured compared with the chance of developing side effects, including a calculation of the total chance of developing side effects.	3–4
Level 1: There is a basic attempt at a comparison between the chances of being cured compared to the chance of developing side effects with no attempt to calculate the total chance of developing side effects.	1–2
No relevant content.	0

Atomic Structure

Nuclear Fission and Fusion

You need to be able to describe the processes of both nuclear **fission** and fusion, including why chain reactions occur.

a) What type of nuclear reaction is shown in the diagram below?
   Explain your answer. *(2 marks)*

$$^{235}_{92}U + ^{1}_{0}n \longrightarrow ^{140}_{56}Ba + ^{93}_{36}Kr + 3^{1}_{0}n$$

Nuclear fission ✓, as the larger atom is splitting into two smaller atoms and emitting neutrons ✓.

b) What other type of nuclear radiation is emitted in this reaction? *(1 mark)*

Gamma rays ✓

c) Explain how the neutrons emitted may go on to cause a change reaction. *(3 marks)*

The neutrons have kinetic energy ✓, they are absorbed by other unstable nuclei ✓, which also undergo fission, emitting more neutrons, which continue the process ✓.

d) Explain how the reaction above differs in a nuclear reactor compared to a nuclear weapon. *(2 marks)*

*In a nuclear reactor the **chain reaction** is controlled to control the energy released ✓. In a nuclear weapon an uncontrolled nuclear reaction occurs, leading to an explosion ✓.*

For more on the topics covered in this chapter, see pages 86–93 of the *Collins GCSE AQA Physics Revision Guide*.

5 Forces

Forces and their Interactions

This topic covers the differences between contact and non-contact forces, **gravity** and resultant forces.

Example

Complete the table with answers from the box below. *(2 marks)*

| friction | tension | magnetic | gravitational | air resistance | electrostatic |

Contact forces	Non-contact forces

Contact forces	Non-contact forces
friction	gravitational
tension	magnetic
air resistance	electrostatic
✓	✓

A **contact force** is a force between objects that have to be physically touching. Non-contact forces can occur between objects that are physically separated.

Example

a) Write down the equation that links weight, mass and gravitational field strength.
(1 mark)

weight = mass × gravitational field strength ✓

The units of this equation are weight in newtons (N), mass in kg and gravitational field strength in N/kg.

Forces

b) Calculate the weight of a person with a mass of 74 kg. Assume gravitational field strength = 9.8 N/kg. *(2 marks)*

> weight = 74 × 9.8 ✓
> weight = 725.2 N ✓

In any calculation involving gravitational field strength the value will be given. The gravitational field strength on Earth is approximately 9.8 N/kg.

c) The gravitational field strength on Mars is approximately 3.8 N/kg.
How would the weight of the person in question **b)** differ if they were on Mars? Explain your answer. *(2 marks)*

> Their weight would be lower ✓. As the gravitational field strength on Mars is lower than on Earth ✓.

Different planets have different gravitational field strength, which means that objects with the same mass would have different weights on different planets.

d) Calculate the mass of an object with a weight of 675 kg on Mars. *(2 marks)*

> $mass = \dfrac{weight}{gravitational\ field\ strength}$
> $mass = \dfrac{675}{3.8}$ ✓
> mass = 177.6 kg ✓

Rearrange the weight equation to find mass and use the gravitational field strength value from part **c)**.

e) What is the single point where the weight of an object acts called? *(1 mark)*

> Centre of mass ✓

Forces

Example

The diagram below shows the forces acting on a moving truck.

⇐ truck ⇒
100 kN   500 kN

a) Calculate the **resultant** force on this truck. *(2 marks)*

500 − 100 ✓
= 400 kN ✓

> In this question the resultant force is found by taking the larger force away from the smaller force. Make sure to use the correct units, which in this case are kN.

b) At a point later in the same journey the forces on the truck changed to those below.

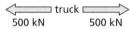
⇐ truck ⇒
500 kN   500 kN

i) What statement can be made about the forces acting on the truck now? Explain your answer. *(2 marks)*

The forces on the truck are balanced ✓, *as the forces are equal, so the resultant force on the truck is zero* ✓.

Work Done and Energy Transfer

This short topic is mostly based around the work done equation. This is an equation you must be able to recall and apply.

Example

a) Write down the equation which relates work done, force and distance. *(1 mark)*

work done = force × distance

> The units of this equation are work done in joules, force in newtons and distance in metres.

b) i) What work is done when a machine uses a force of 459 N to move object A 7 m? *(2 marks)*

work done = 459 × 7 ✓
work done = 3213 J ✓

Forces

ii) The settings of the machine are changed to move object B. The work done to move object B 4.2 m is 4918 J.
What is the force exerted on object B by the machine? (2 marks)

> force = $\frac{\text{work done}}{\text{distance}}$
> force = $\frac{4918}{4.2}$ ✓
> force = 1171 N ✓

Rearrange the equation from part **a)** so that force is the subject of the equation.

iii) Whilst object B is moving its temperature increases.
Explain why. (2 marks)

> Frictional forces act on the object as it moves ✓. Work done against these forces causes a rise in temperature ✓.

Unless an object is in a vacuum, friction will always be produced when an object moves.

c) What is 1318 Nm in kJ?
Give your answer to two significant figures. (1 mark)

> 1318 Nm = 1.3 kJ ✓

1 joule (J) = 1 newton-metre (Nm). An answer to two significant figures should contain two digits that are not leading or trailing zeroes (zeroes at the start of the number).

Forces and Elasticity

This topic covers the effects of stretching, bending or compressing leading to elastic deformation or inelastic deformation.

Example

a) Write down the equation that links force applied to a spring, **spring constant** and extension. (1 mark)

> force applied to a spring = spring constant × extension ✓

The units of this equation are force in newtons (N), spring constant in newtons per metre (N/m) and extension in metres (m).

Forces

a) What force is needed to extend a spring 0.36 m where the spring has a spring constant of 65 N/m? *(2 marks)*

$force = 65 \times 0.36$ ✓
$force = 23.4 \, N$ ✓

b) The same force was used to compress a large rubber ball which has a spring constant of 110 N/m.
What is the **compression** of the rubber ball? *(2 marks)*

$extension \, (compression) = \dfrac{force}{spring \, constant}$

$compression = \dfrac{23.4}{110}$ ✓

$compression = 0.21 \, m$ ✓

The equation from part **a)** can be used to calculate compression, where extension is the compression value. Rearrange the equation from **a)** to find the extension.

Example

The graphs below show the force over extension for two objects.

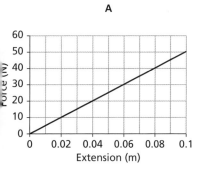
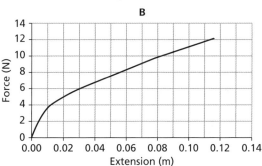

c) Which of the graphs show a linear relationship? *(1 mark)*

Graph A ✓

Extension of an elastic object is directly proportional to the force applied.

Forces

b) Calculate the spring constant of the object in graph A. *(3 marks)*

> Values from graph:
> force = 10 N
> extension = 0.02 m ✓
> spring constant = $\frac{force}{extension}$
> spring constant = $\frac{10}{0.02}$ ✓
> spring constant = 500 N/m ✓

> To calculate the spring constant take a force and extension value from the graph and rearrange the force extension equation to find the spring constant.

c) Would graph A continue with the same relationship if the force applied caused the limit of proportionality to be exceeded?
Explain your answer. *(2 marks)*

> No, it would no longer be a linear relationship ✓, as force and extension will no longer be proportional ✓.

d) What is the elastic potential energy of the object at 0.06 m on Graph A?
Show how you arrived at your answer. *(2 marks)*

> elastic potential energy = 0.5 × spring constant × (extension)^2
> elastic potential energy = 0.5 × 500 × 0.06^2 ✓
> elastic potential energy = 0.9 N ✓

> This question requires you to recall and apply the equation for elastic potential energy from the Energy topic.

e) What is the work done to extend the object?
Explain your answer. *(2 marks)*

> Work done = 0.9 N ✓. This is because the elastic potential energy stored is equal to work done if the spring is not **inelastically deformed** ✓.

Forces

Moments, Levers and Gears

The turning effect of a force is called the **moment** of the force. **Levers** and **gears** can be used to transmit the rotational effect of forces.

Example

a) Write down the equation that links moment of a force to force and distance. *(1 mark)*

> moment of a force = force × distance ✓

The units in this equation are moment of a force in newton-metres, force in newtons and distance in metres. The distance is the perpendicular distance from the pivot to the line of action.

b) What is the moment of a force applied by a lever when distance from the **pivot** is 0.6 m and the force applied to the lever is 270 N? *(2 marks)*

> moment of a force = 270 × 0.6 ✓
> moment of a force = 162 Nm ✓

c) The force on the lever increases, producing a moment of 205 Nm. What is the new force? *(2 marks)*

> $force = \frac{moment\ of\ a\ force}{distance}$
> $force = \frac{205}{0.6}$ ✓
> force = 342 Nm ✓

To find force, rearrange the equation from a) to make force the subject of the equation and use the same distance from the pivot.

Example

The diagram below shows two objects on a see-saw. The see-saw is balanced.

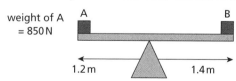

Forces

a) What statement can be made about the clockwise moment and anticlockwise moment around the pivot? *(1 mark)*

   The clockwise moment around the pivot is equal to the anticlockwise moment around the pivot ✓.

b) Calculate the mass of object B; assume gravitational field strength = 9.8 N/kg. Write down any equations you use. *(6 marks)*

   moment of object A = force × distance
   moment of object A = 850 N × 1.2 ✓
   moment of object A = 1020 Nm ✓
   moment of object B = 1020 Nm
   $force = \frac{moment}{distance}$
   $force = \frac{1020}{1.4}$ ✓
   force = 729 N ✓
   weight of object B = 729 N
   $mass = \frac{weight}{gravitational\ field\ strength}$
   $mass = \frac{729}{9.8}$ ✓
   mass = 74 kg ✓

   This is a challenging question that requires you to link different areas of the Forces topic together.
   1. First calculate the moment of object B; this is equal to the moment of object A.
   2. Next rearrange the moment equation to find the force produced by object B.
   3. Finally, rearrange the weight equation to calculate the mass of object B.

Pressure and Pressure Differences in Fluids

This topic covers two equations used when determining pressure in fluids, and a simple model of atmospheric pressure.

Example

Which of the following is not a **fluid**? Tick **one** box. *(1 mark)*

liquid ☐
solid ☐
gas ☐

solid ✓

Forces

Example

A scuba diver is taking part in a dive trip to a coral reef.

a) The diver submerges her mask in water to rinse it off. Whilst doing this a force of 15 N is exerted on the screen of the mask, which has an area of 0.02 m².

  i) Write down the equation that links pressure, force and area. *(1 mark)*

  $$pressure = \frac{force\ normal\ to\ a\ surface}{area\ of\ that\ surface}\ \checkmark$$

  > The units of pressure are pascals (Pa), area in m² and force in newtons (N).

  ii) What is the pressure applied to the mask screen? *(2 marks)*

  $$pressure = \frac{15}{0.02}\ \checkmark$$
  $$pressure = 750\ Pa\ \checkmark$$

  iii) The diver's dive partner does the same with his mask. He exerts a force of 17 N and applies a pressure of 680 Pa to the mask screen.
  What is the area of his mask screen? *(2 marks)*

  $$area = \frac{force}{pressure}$$
  $$area = \frac{17}{680}\ \checkmark$$
  $$area = 0.025\ m^2\ \checkmark$$

  > As the pressure is lower than in part ii), even though the force applied is greater, this means the area of the mask must be greater than that in part ii).

b) i) **HT** The divers descend to a depth of 18 m. What is the pressure due to the column of liquid at this point?
  The density of seawater is 1020 kg/m³ and the gravitational field strength is 9.8 N/kg.
  Use the correct equation from the Physics Equations Sheet and give your answer in KPa to three significant figures. *(2 marks)*

  $$pressure = height\ of\ the\ column \times density\ of\ the\ liquid \times gravitational\ field\ strength$$
  $$pressure = 18.0 \times 1020 \times 9.8\ \checkmark$$
  $$pressure = 179\,928\ Pa = 180\ kPa\ \checkmark$$

Forces

ii) **HT** If the divers descended to the same depth in fresh water the pressure would be 176.4 kPa.
Use this information to calculate the density of fresh water. *(2 marks)*

density of the liquid = pressure ÷ height of the column ÷ gravitational field strength
density of the liquid = 176 400 ÷ 18 ÷ 9.8 ✓
density of the liquid = 1000 kg/m³ ✓

> To find the density of the water, rearrange the equation to make the density of the liquid the subject. Make sure to convert 176.4 kPa into Pa before substituting into the equation.

c) i) After observing the coral the divers return to their dive boat.
Explain in terms of forces why the dive boat floats. *(2 marks)*

The boat experiences a greater pressure on its bottom surface than on the top surface ✓. This creates a resultant force upwards known as **upthrust** ✓.

ii) If the dive boat was overloaded it would begin to sink.
Explain why. *(2 marks)*

The boat has become more dense than the surrounding liquid ✓, so is unable to **displace** a volume of liquid equal to its own weight. The boat sinks as its weight is greater than the upthrust ✓.

Example

A plane takes off from Heathrow on a flight to New York. The plane ascends to a height of 12 000 m.

a) What creates **atmospheric pressure**? *(1 mark)*

Air molecules colliding with a surface ✓.

b) How would the atmospheric pressure around the plane change during this time? Explain your answer. *(2 marks)*

It would decrease ✓; as the plane gains height the number of air molecules above the plane decrease, so decreasing the atmospheric pressure ✓.

Forces

Forces and Motion

This is a wide-ranging topic covering speed, velocity, acceleration and Newton's Laws of motion. Forces and braking covers the factors that affect the ability of a vehicle to come to a stop.

Example

a) Match the different activities to the typical speeds. *(2 marks)*

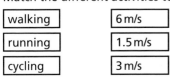

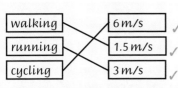

−1 mark for each incorrect answer.

b) A sound wave has a speed of 330 m/s in air.

  i) Write down the equation that links distance, speed and time. *(1 mark)*

  distance travelled = speed × time ✓

  > The units in this equation are distance in metres (m), speed in m/s and time in seconds (s).

  ii) How far would the sound wave travel in 15 seconds? *(2 marks)*

  distance travelled = 330 × 15 ✓
  distance = 4950 m ✓

  iii) The same sound wave would travel 22 km in water in the same time. What is the speed of sound in water? *(2 marks)*

  speed = $\frac{distance}{time}$
  speed = $\frac{22000}{15}$ ✓
  speed = 1467 m/s ✓

  > Rearrange the equation so that speed is the subject of the equation. Make sure to convert 22 km into metres before substituting into the equation.

Forces

Example

A car accelerates from 15 m/s to 27 m/s in 4 seconds.

a) Write down the equation that links **acceleration**, change in **velocity** and time taken. *(1 mark)*

$$\text{acceleration} = \frac{\text{change in velocity}}{\text{time taken}} \checkmark$$

> The units of this equation are acceleration in m/s², change in velocity in m/s and time taken in s.

b) Calculate the acceleration of the car. *(2 marks)*

$$\text{acceleration} = \frac{(27 - 15)}{4}$$
$$\text{acceleration} = \frac{12}{4} \checkmark$$
$$\text{acceleration} = 3 \, \text{m/s}^2 \checkmark$$

> To find the change in velocity take the final velocity away from the initial velocity.

c) The car changes speed from 29 m/s to 15 m/s with a **deceleration** of 2 m/s. What time did it take for this deceleration to occur? *(2 marks)*

$$\text{time taken} = \frac{\text{change in velocity}}{\text{acceleration}}$$
$$\text{time taken} = \frac{(15 - 29)}{-2} \checkmark$$
$$\text{time taken} = \frac{-14}{-2}$$
$$\text{time taken} = 7 \, \text{seconds} \checkmark$$

> As the car is going from a higher velocity to a lower velocity this is a deceleration, so will have a negative number for change in velocity.

Forces

Example

The graph below shows the motion of boat.

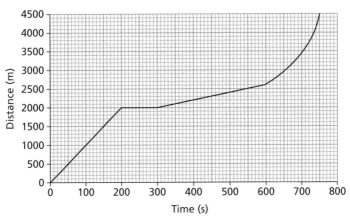

a) At what points is the boat stationary?
Explain your answer. *(2 marks)*

From 200 s to 300 s ✓, as the line is flat so the distance travelled by the boat is constant at this point ✓.

b) In which of the following time periods is the boat travelling the fastest?
Explain how you arrived at your answer. *(2 marks)*

0–200 s

300–600 s

0–200 s ✓, as this is the point where the line is the steepest ✓.

> The speed of an object can be determined by the **gradient** of its distance–time graph.

c) What is the speed of the boat from 0 to 100?
Show your working. *(3 marks)*

change in distance = 1000 − 0 = 1000 m travelled

change in time = 100 − 0 = 100 s ✓

speed = $\frac{distance}{time}$

speed = $\frac{1000}{100}$ ✓

speed = 10 m/s ✓

Forces

Calculate the gradient of the line by dividing the change in the variable on the y-axis (in this case distance) by the corresponding change in the variable on the x-axis (in this case time).

d) **HT** How could you determine the speed at 700 s? *(2 marks)*

Draw a tangent to the curve at 700 s ✓ *and find the gradient of the tangent. This will be the speed of the boat* ✓.

The line is curving upwards at 700 s so the boat is accelerating.

Example

a) Explain why speed is a **scalar** quantity whilst velocity is a **vector** quantity. *(2 marks)*

Speed only has magnitude so is a scalar quantity ✓. *Velocity has a magnitude and a direction so is a vector quantity* ✓.

b) A motorbike accelerates from 0 to 30 m/s in 6 s.
Use the correct equations from the Physics Equations Sheet to calculate the distance it covers in this time. *(5 marks)*

$$(\text{final velocity})^2 - (\text{initial velocity})^2 = 2 \times \text{acceleration} \times \text{distance}$$

First, calculate the acceleration of the motorbike:

$$\text{acceleration} = \frac{\text{change in velocity}}{\text{time taken}}$$

$$\text{acceleration} = \frac{(30-0)}{6}$$

$$\text{acceleration} = \frac{30}{6} \checkmark$$

$$\text{acceleration} = 5 \, \text{m/s}^2 \checkmark$$

Now find the result of the left side of the equation:

$$(\text{final velocity})^2 - (\text{initial velocity})^2 = 30^2 - 0^2$$

$$(\text{final velocity})^2 - (\text{initial velocity})^2 = 900 \checkmark$$

$$900 = 2 \times \text{acceleration} \times \text{distance}$$

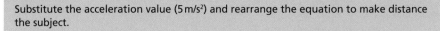

Substitute the acceleration value (5 m/s²) and rearrange the equation to make distance the subject.

$distance = 900 \div 2 \div 5$ ✓

$distance = 90 \, m$ ✓

Example

The graph below shows the velocity of a runner during a race.

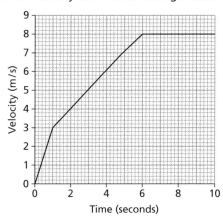

a) What was the runner's acceleration between 2 and 6 seconds? *(2 marks)*

$change\ in\ velocity = 8 - 4 = 4\ s$ ✓

$change\ in\ time = 6 - 2 = 4\ s$

$acceleration = \frac{4}{4} = 1\ m/s^2$ ✓

The acceleration of an object can be calculated from the gradient of a velocity–time graph.

b) **HT** What distance does the runner travel between 6 and 10 s? *(2 marks)*

$area\ under\ graph\ from\ 6\ to\ 10\ s = (10 - 6) \times (8 - 0)$

$area\ under\ graph\ from\ 6\ to\ 10\ s = 4 \times 8$ ✓

$area\ under\ graph\ from\ 6\ to\ 10\ s = 32\ m$ ✓

The distance travelled by an object (or **displacement** of an object) can be calculated from the area under a velocity–time graph.

Forces

c) i) What distance does the runner run between 6 and 2 s? (2 marks)

> area under graph from 2 to 6 s = 0.5 × (6−2) × (8−4) + (6−2) × (4−0) ✓
> area under graph from 2 to 6 s = 8 + 16
> area under graph from 2 to 6 s = 24m ✓

ii) Explain the difference in your answers to **b)** and **c) i)**. (2 marks)

> As the runner's velocity was greater at 6–10 s than at 2–4 s ✓,
> the distance covered in the same length of time (4 s) was greater ✓.

Forces, Accelerations and Newton's Laws of Motion

Example

Use the answers in the box to match up Newton's Laws of motion with the correct description. (2 marks)

| First Law | Second Law | Third Law |

Description	Law
If the resultant force acting on an object is zero and the object is stationary, the object remains stationary.	
Whenever two objects interact, the forces they exert on each other are equal and opposite.	
The acceleration of an object is proportional to the resultant force acting on the object.	

Description	Law
If the resultant force acting on an object is zero and the object is stationary, the object remains stationary.	First Law
Whenever two objects interact, the forces they exert on each other are equal and opposite.	Third Law
The acceleration of an object is proportional to the resultant force acting on the object.	Second Law

✓ ✓ One incorrect answer −1 mark.

Forces

Example

A cyclist is riding down a mountain road.

a) The cyclist is maintaining a constant velocity.
What statement can be made about the restive force and the driving force of the cyclist? *(1 mark)*

They are balanced ✓.

b) The cyclist begins pedalling faster and accelerates.
What statement can be made about the restive force and driving force now?
Use the term 'resultant force' in your answer. *(2 marks)*

The driving force is greater than the resistive force ✓, *this means that there is now a resultant force on the cyclist* ✓.

c) **HT** The cyclist comes to a stop and will remain in this state of rest until she accelerates again. What term is given to the tendency of objects to continue in their state of rest? *(1 mark)*

Inertia ✓

Example

A skier (A) who has a mass of 90 kg accelerates at 3 m/s².

a) Write down the equation that links resultant force, mass and acceleration. *(1 mark)*

resultant force = mass × acceleration ✓

The units of this equation are force in N, mass in kg and acceleration in m/s².

b) What is the resultant force on the skier? *(2 marks)*

resultant force = 90 × 3 ✓
resultant force = 270 N ✓

c) A second skier (B) accelerates at the same rate but has a resultant force acting on them of 240 N. What is their mass? *(2 marks)*

$$mass = \frac{resultant\ force}{acceleration}$$
$$mass = \frac{240}{3}\ ✓$$
$$mass = 80\ kg\ ✓$$

As the resultant force is lower, this means the mass of skier B must be lower than that of the first skier.

Forces

HT d) If the skiers are travelling at the same speed, skier A will have the greater inertial mass.
What is meant by the term inertial mass? *(1 mark)*

> Inertial mass is a measure of how difficult it is to change the velocity of an object ✓.

Example

a) What **two** distances make up the stopping distance? *(1 mark)*

> The thinking distance and the **braking distance** ✓.

> The thinking distance is the distance the vehicle travels during the driver's reaction time. The braking distance is the distance the vehicle travels under the braking force.

b) Which of the below values is a typical human reaction time? Tick **one** box. *(1 mark)*

0.02 seconds ☐
0.2 seconds ☐
2 seconds ☐

> 0.2 seconds ✓

c) Which of the following factors would **not** affect the thinking distance?
Tick **two** boxes. *(1 mark)*

The condition of the tyres ☐
The driver having consumed alcohol ☐
The driver being tired ☐
A wet road ☐

> The condition of the tyres; A wet road ✓

Forces

1) On French motorways the speed limit is 130 km/h in the dry and 110 km/h when it is raining. Use the idea of stopping distance to explain why this is the case. *(3 marks)*

> *On a wet road stopping distance increases due to the braking distance increasing ✓. Speed also affects stopping distance so a car travelling at 130 km/h would have a much longer stopping distance on a wet road than a car travelling at 110 km/h ✓. As the stopping distance is shorter, travelling at 110 km/h reduces the chance of accidents occurring ✓.*

> You should relate your answer to the specific context of the question, including using the **two** speed values in your answer.

Example

2) Explain why a vehicle stops when the brakes are applied. *(3 marks)*

> *When a force is applied to the brakes, work done ✓ by the friction force between the brakes and the wheel ✓ reduces the kinetic energy of the vehicle until it reaches zero and the vehicle stops ✓.*

3) Many high-performance cars have specially designed air-flow systems to maximise the flow of air over the brakes. This is particularly important during large decelerations.
Explain why. *(4 marks)*

> *During large decelerations, work done by the friction force of the brakes against the wheels ✓ leads to large increases in the temperature of the brakes ✓. This can lead to the brakes overheating and stop working ✓. By maximising the air flow over the brakes they will be cooled down during large decelerations ✓.*

Example

Reaction time can be measured using an app on a smartphone. When the colour of the screen changes the testee must press a button as quickly as possible. The time it takes them to press the button is then displayed by the app.

Design an investigation into the effect of distraction on thinking distance. Suggest methods to ensure the results are not seriously affected by random error and **anomalous** results are correctly dealt with. *(6 marks)*

Forces

First have the subject complete the test multiple times; this will allow them to get used to the test and reduce the effect of learning on the results. Once their reaction times are consistent, begin recording the times. Make sure they do the test in a silent room with no distractions. After a set number of tests, e.g. 20, give the subject a short break. The subject should now repeat the test the same number of times but during each test they should be distracted in the same way by being talked to. Record their reaction time after each test. Once you have completed your investigation discard any results that are very different from the others (anomalous results). Calculate a mean reaction time for completing the test undistracted and a mean reaction time for completing the test whilst being distracted.

This is an example of a 6-mark extended response question. It is marked in levels. The table below shows the different level descriptors. In order for your answer to be placed in a level you have to satisfy the criteria laid out for that particular level. The quality of your answer then determines what mark you are awarded within that level. The key aspect of all these questions is developing a sustained line of reasoning which is coherent, relevant, substantiated and logically structured. This means you should focus on writing well structured answers with a logical order that relate directly to the question and do not contain any non-relevant material.

When designing an investigation in an exam ensure your instructions are clear and logically ordered, your method ensures the experiment is **reproducible** and you've taken steps to reduce random and systematic errors.

Level	Marks
Level 3: Clear and coherent description of investigation method including correct description of discarding anomalous results and calculation of a mean.	5–6
Level 2: Partial description of investigation, which may not be logically ordered. Correct reference made to anomalous results but lacking in a clear explanation of how to deal with them.	3–4
Level 1: Basic description of simple method with incorrect or very underdeveloped reference to anomalous results.	1–2
No relevant content.	0

Forces

ⓗ Momentum

Students often find **momentum** a challenging topic – some of the calculations can become quite complex – but the key thing to remember is that in a closed system total momentum before an event is equal to total momentum after an event. This is conservation of momentum.

Example

A car with a mass of 1300 kg is travelling at 15 m/s during a crash test.

a) Write down the equation that relates momentum, mass and velocity. *(1 mark)*

> momentum = mass × velocity ✓
>
> The units for this equation are momentum in kg m/s, mass in kg and velocity in m/s.

b) Calculate the momentum of the car. *(2 marks)*

> momentum = 1300 × 15 ✓
> momentum = 19 500 kg m/s ✓

c) The car changed velocity and its momentum changed to 24 700 kg m/s. What was the car's new velocity? *(2 marks)*

> $velocity = \frac{momentum}{mass}$
> $velocity = \frac{24\,700}{1300}$ ✓
> velocity = 19 m/s ✓
>
> Rearrange the equation from part a) so that velocity is the subject.

d) The car hits another car, which is stationary. What is the total momentum of the two cars when they collide?
Explain your answer. *(3 marks)*

> Momentum is conserved in a **collision** ✓. The momentum of the moving car is 24 700 kg m/s. As the other car is stationary, its momentum is 0 ✓. Total momentum = 24 700 + 0 = 24 700 kg m/s ✓.

Forces

e) A crash test dummy (dummy A) in the moving car has a mass of 85 kg and comes to rest in 0.2 seconds.
Calculate the force it experiences.
Give your answer in kN and write down any equations you use. *(4 marks)*

> $momentum = mass \times velocity$
> $momentum = 85 \times 19$ ✓
> $momentum = 1615 \text{ kg m/s}$ ✓
> $force = \dfrac{change\ in\ momentum}{time\ taken}$
> $force = \dfrac{(1615 - 0)}{0.2}$ ✓
> $force = \dfrac{1615}{0.2}$
> $force = 8075 \text{ N}$ ✓

> First use the equation in part **a)** to calculate the dummy's momentum. The equation which links force, change in momentum and time taken is given on the Physics Equations Sheet.

f) In a repeat of this test, with the car travelling at the same velocity, the force experienced by a dummy of the same mass (dummy B) was 107 kN.
What was the time taken for the change in momentum? *(2 marks)*

> $time\ taken = \dfrac{change\ in\ momentum}{force}$
> $time\ taken = \dfrac{1615}{107000}$ ✓
> $time\ taken = 0.015 \text{ seconds}$ ✓

> As both dummies have the same mass and were travelling at the same velocity, their momentums will be the same. The force is given in kN so convert it into newtons before substituting it into the rearranged equation.

Forces

g) Which of the dummies was wearing a seatbelt?
Explain how you arrived at your answer. *(2 marks)*

> A seatbelt works by increasing the time it takes for a body to come to rest ✓.
> Dummy A took the longest time to come to rest, so was wearing a seatbelt ✓.

A seatbelt is an example of a safety device that increases the time taken for a change in momentum to occur.

h) By comparing your answers to **e)** and **f)** explain the importance of wearing seatbelts when travelling in a car. *(3 marks)*

> By increasing the time taken for the change in momentum to occur, the seatbelt significantly reduces the force on the dummies ✓ from 107 kN to 8075 N, which is over 100 times less force ✓. This significantly reduces the risk of injury ✓.

The key command word in this question is 'compare'. Make sure you include references to both **e)** and **f)** in your answer.

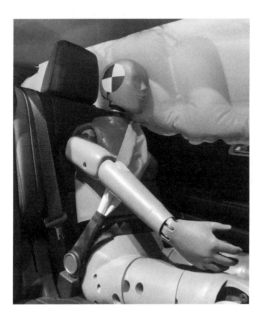

For more on the topics covered in this chapter, see pages 8–21 of the *Collins GCSE AQA Physics Revision Guide*.

6 Waves

Waves in Air, Fluids and Solids

This topic looks at a number of different properties of waves, including reflection, sound waves and how waves can be used for detection and exploration of structures that are hidden from direct observation.

Example

a) Use the answers in the box below to complete the table on the features of a wave. *(2 marks)*

| wavelength | amplitude | frequency |

Definition	Key feature
The number of waves passing a point each second.	
The maximum displacement of a point on a wave away from its undisturbed position.	
The distance from a point on one wave to the equivalent point on the adjacent wave.	

Definition	Key feature
The number of waves passing a point each second.	frequency
The maximum displacement of a point on a wave away from its undisturbed position.	amplitude
The distance from a point on one wave to the equivalent point on the adjacent wave.	wavelength

✓✓ One incorrect answer −1 mark.

Waves

b) Which of the following statements is not a feature of **longitudinal waves**?
Tick **one** box. *(1 mark)*

Show areas of compression and rarefaction. ☐

The **oscillations** of the wave are perpendicular to the direction of energy transfer. ☐

The oscillations of the wave are parallel to the direction of energy transfer. ☐

The oscillations of the wave are perpendicular to the direction of energy transfer. ✓

Example

A group of school students were carrying out an investigation into waves. They were using a ripple tank and a stroboscope.

a) What type of waves were they investigating? *(1 mark)*

The students were investigating water waves ✓.

b) The students placed a ping-pong ball in the tank. The ball floated and moved up and down as the wave passed, but did not move along the tank. Explain the property of waves that this provides evidence of. *(2 marks)*

As the ball only moved up and down when the water wave passed, this shows that the waves transfer energy and information ✓ *but do not transfer matter* ✓.

Example

a) The frequency of a wave is 250 Hz. What is the **period** of this wave?
Write down any equations you use. *(2 marks)*

$period = \frac{1}{frequency}$ ✓
$period = \frac{1}{250}$
$period = 0.004 \text{ s}$ ✓

b) Write down the equation that links wave speed, frequency and wavelength.
(1 mark)

wave speed = frequency × wavelength ✓

> The units of this equation are wave speed in m/s, frequency in Hz and wavelength in m.

Waves

c) What is the speed of a wave that had a frequency of 3 kHz and a wavelength of 0.03 m? *(2 marks)*

> wave speed = 3000 × 0.03 ✓
> wave speed = 90 m/s ✓

Convert 3 kHz into Hz before substituting the value into the equation. There are 1000 Hz in 1 kHz.

d) The frequency of the wave changes but the wavelength doesn't alter. The new wave speed is 2.7×10^3 m/s.
What is the frequency of the wave?
Give your answer in kHz. *(2 marks)*

> $frequency = \dfrac{wave\ speed}{wavelength}$
>
> $frequency = \dfrac{2700}{0.03}$ ✓
>
> $frequency = 90$ kHz ✓

The wave speed has been given in standard form – as it is to the power ×10³ you need to move the decimal point back three places. This gives a value of 2700 m/s.

Example

The diagram below shows the reflection of a light ray.

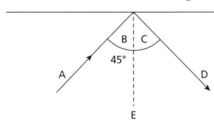

a) Identify A to E. *(5 marks)*

> A = incident ray ✓
> B = angle of incidence ✓
> C = angle of reflection ✓
> D = **reflected** ray ✓
> E = **normal** ✓

Waves

b) What is the value for C?
Explain how you arrived at your answer. *(2 marks)*

> 45° ✓ because the angle of incidence is equal to the angle of reflection ✓.

c) In this example total internal reflection has occurred.
What can be concluded about the critical angle?
Explain how you arrived at your answer. *(2 marks)*

> The critical angle must be less than 45° ✓, as the angle of incidence is 45° and total internal reflection only occurs if the angle of incidence is greater than the critical angle ✓.

HT Example

The table below shows the hearing ranges of different animals.

Animal	Lower hearing limit (Hz)	Upper hearing limit (kHz)
cat	55	79
bat	15 000	90
mouse	1000	70
human		

a) Complete the table with the hearing range of humans. *(2 marks)*

human	20	20

✓ ✓ One incorrect answer –1 mark.

b) Which animal would be able to hear a sound with a frequency of 8.5×10^4 Hz? *(1 mark)*

> bat ✓
>
> This value is equivalent to 85 kHz.

Waves

c) Which of the animals can hear **ultrasound**?
Explain your answer. (2 marks)

> All the animals can hear ultrasound except humans ✓. Ultrasound is sound with a frequency greater than 20 kHz and all the animals have hearing ranges that go above 20 kHz ✓.

HT Example

A common injury that scuba divers suffer from is perforated eardrums. When an eardrum perforates it tears. One of the symptoms of perforated eardrums is loss of hearing.
Explain why. (3 marks)

> Sound waves cause the eardrum to vibrate ✓ and these vibrations are changed into signals that pass to the brain ✓. If the eardrum is torn it will be unable to vibrate properly, which will mean that the vibrations will not be properly passed to the brain to be perceived as sounds ✓.

HT Example

The image below shows a prenatal scan.

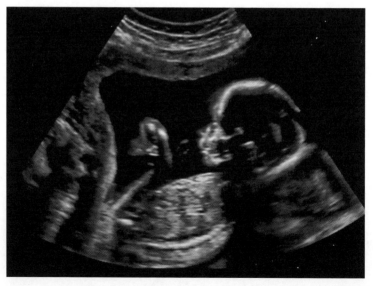

Waves

Explain how the image was produced. *(3 marks)*

> By ultrasound waves ✓. Ultrasound waves are partially reflected when they meet a boundary between two different media ✓. The time taken for the reflections to reach a detector can be used to determine how far away the boundary is ✓. This information is then used to produce an image of these different boundaries.

HT Example

The diagram below shows seismograph traces produced by a minor earthquake which took place in New Zealand. Detector A was at a site in Portugal, which is on the opposite side of the Earth to the epicentre. Detector B was in Australia, approx. 1300 miles from the epicentre of the earthquake.

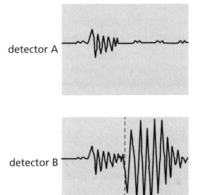

X and Y represent two different types of waves.
Identify waves X and Y.
Explain how you arrived at your answer.
You should relate your answer to the structure of the Earth. *(6 marks)*

X: **P-waves** ✓

Y: **S-waves** ✓

P-waves are longitudinal **seismic** waves that can travel through solids and liquids ✓. S-waves are **transverse** seismic waves that can travel through solids but not liquids ✓. The outer core of the Earth's surface is liquid, which means that the P-waves pass through and are detected on the opposite side of the Earth at A, whilst S-waves cannot be detected at A ✓. Both kinds of waves are able to reach B so both are detected there ✓.

Waves

Example

A fishing boat was tracking a shoal of fish. The boat released a wave and 0.65 seconds later detected the wave's reflection off the fish shoal.

a) What type of wave was released by the fishing boat? *(1 mark)*

> Sound ✓

b) The speed of the wave in water was 1500 m/s.
How far below the boat was the shoal of fish?
Write down any equations you use. *(3 marks)*

> distance = speed × time
> distance = 1500 × 0.65 ✓
> distance = 975 m ✓
> $\frac{975}{2}$ = 487.5 m ✓

> Once you've found the distance, divide this value by 2. This represents the fact that the wave has to travel from the boat to the shoal and then back to the boat again.

Electromagnetic Waves

Electromagnetic waves are transverse waves that transfer energy from the source of the waves to an absorber. Electromagnetic waves form a continuous spectrum and all types of electromagnetic wave travel at the same velocity through a vacuum or air.

Example

a) Complete the electromagnetic spectrum using the answers in the box below.
(2 marks)

| microwaves | ultraviolet | radio waves |

long wavelength ⟶ short wavelength

| A | B | infrared | visible light | C | X-rays | gamma rays |

A = radio waves   B = microwaves   C = ultraviolet ✓ ✓

One incorrect answer −1 mark.

Waves

b) Which type of electromagnetic radiation is detected by our eyes? *(1 mark)*

Visible light ✓

Example

The diagram below shows the refraction of a light ray as it travels between two media.

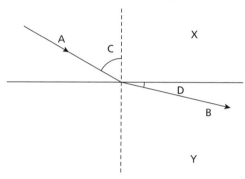

a) Identify A to D. *(4 marks)*

A = incident ray ✓
B = **refracted** ray ✓
C = angle of incidence ✓
D = angle of refraction ✓

b) **HT** Which of the media (X or Y) is the most optically **dense**?
Explain how you arrived at your answer. *(2 marks)*

Media X is the most dense ✓, as the light ray bent away from the normal has moved from media X to media Y ✓.

Light bends towards the normal when entering a more optically dense **medium** and bends away from the normal when entering a less optically dense medium.

Waves

c) **HT** The diagram below shows a wave moving between air and water. Use the diagram to explain why the wave has refracted. *(3 marks)*

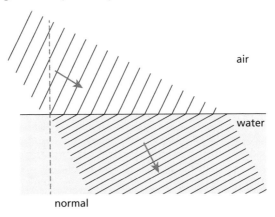

As the wave travels slower in a denser medium ✓, the edge of the wave that hits the water first slows down while the rest of the wave continues at the same speed ✓. This causes the wave to bend towards the normal ✓.

Example

In an experiment to investigate radio waves an antenna was set up to detect the radio waves.

a) **HT** Explain how radio waves are produced. *(1 mark)*

Radio waves can be produced by oscillations in electrical circuits ✓.

b) **HT** The antenna is connected to a circuit that can measure the frequency of an alternating current.
Explain how this apparatus can be used to measure the frequency of radio waves. *(2 marks)*

When radio waves are absorbed they may create an alternating current ✓ with the same frequency as the radio wave ✓.

Waves

c) Use the words in the box below to complete the following passage. Each word can be used once, more than once, or not at all. (2 marks)

| ultraviolet | gamma | waves | nucleus | atoms |

Changes in _____ and the nuclei of atoms can result in electromagnetic _____ being generated or absorbed over a wide frequency range. _____ rays originate from changes in the _____ of an atom.

Changes in **atoms** and the nuclei of atoms can result in electromagnetic **waves** being generated or absorbed over a wide frequency range. **Gamma** rays originate from changes in the **nucleus** of an atom. ✓ ✓
One incorrect answer –1 mark.

Example

The ray diagram below shows the effect of a lens.

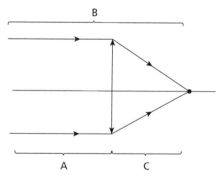

a) Identify the type of lens shown. (1 mark)

This is a convex lens ✓.

There are two types of lenses, **convex** lenses and **concave** lenses.

Waves

b) Which of the distances is the **focal length**?
Explain your answer. *(2 marks)*

> C ✓. The focal length is the distance between the lens and the **principal focus** ✓.

c) The image formed by this lens is a **real** image. How would the image differ if the lens was replaced with the lens shown below?
Explain your answer. *(2 marks)*

> This is a concave lens ✓ so would form a **virtual** image ✓.

Example

a) Explain why **magnification** has no units. *(1 mark)*

> Magnification is a ratio ✓.

b) An image with a height of 400 cm is produced by a lens. The object being magnified has a height of 2.4 mm.

  i) Calculate the magnification.
     Use the correct equation from the Physics Equations Sheet. *(2 marks)*

> 2.4 mm = 0.24 cm
> $$\text{magnification} = \frac{\text{image height}}{\text{object height}}$$
> $$\text{magnification} = \frac{400}{0.24} \checkmark$$
> $$\text{magnification} = 1667 \checkmark$$
>
> Before substituting the values into the equation convert them into the same units (1 cm = 10 mm).

ii) The same object was magnified 2150 times. What is the image size?
Give your answer in metres to two significant figures. *(2 marks)*

> image height = magnification × object height
> image height = 2150 × 0.24 ✓
> image height = 516 cm = 0.52 m ✓

Ensure your final answer is given in metres (100 cm = 1 m).

Example

An investigation was carried out into colours. A red cube was used. The red cube does not transmit any light that falls on it.

a) What term can be used for the cube? *(1 mark)*

> Opaque ✓

Transparent objects transmit light through them whilst opaque objects absorb or reflect all light incident on them.

b) Explain why this cube appears red. *(1 mark)*

> The cube absorbs all wavelengths of light except red light, which is reflected ✓.

c) When the cube is viewed through a blue filter it appears to be black.
Explain why. *(2 marks)*

> The blue filter only allows blue light through ✓. This means that the red light reflected from the cube is not **transmitted**, so the cube appears black ✓.

d) One side of the cube is mirrored, the other sides are not.
Compare the reflection from the different sides of the cube. *(2 marks)*

> **Specular reflection** is occurring from the mirrored side of the cube ✓.
> Diffuse reflection is occurring from the non-mirrored side ✓.

There are **two** types of reflection; specular or diffuse.

Waves

Black Body Radiation

This is a short topic that covers the emission and absorption of **infrared** radiation and the theoretical perfect black bodies.

Example

Complete the passage using the words in the box below. *(2 marks)*

Each word can be used once, more than once, or not at all.

| absorb | temperature | transmit | radiates |

All bodies (objects), no matter what _____, emit and _____ infrared radiation. The hotter the body, the more infrared radiation it _____ in a given time.

> All bodies (objects), no matter what **_temperature_**, emit and **_absorb_** infrared radiation. The hotter the body, the more infrared radiation it **_radiates_** in a given time. ✓✓
> One incorrect answer −1 mark.

HT Example

An investigation was carried out into radiation on a black metal sphere. A source of radiation is incident on the black sphere, and the intensity of this source changes over time.

The results of the investigation are shown in the table below.

Time (minutes)	Temperature of sphere (°C)
0	30
10	30
20	1200
30	700

Waves

a) The black sphere is not a perfect **black body**.
   Give **one** reason why. *(1 mark)*

   > The sphere doesn't absorb all of the radiation incident on it.

   A perfect black body is a theoretical object and is not possible in reality.

b) Explain the results during the following time periods in terms of absorbing and emitting radiation.

   i) 0–10 minutes *(1 mark)*

   > The temperature is constant as the rate of radiation absorbed is equal to the rate of radiation emitted ✓.

   ii) 10–20 minutes *(1 mark)*

   > The temperature is increasing as the rate of radiation absorbed is greater than the rate of radiation emitted ✓.

   iii) 20–30 minutes *(1 mark)*

   > The temperature is decreasing as the rate of radiation absorbed is less than the rate of radiation emitted ✓.

c) At 20 minutes the metal was glowing bright yellow. After 30 minutes it was glowing red and less brightly.
   Explain these observations. *(3 marks)*

   > The intensity of radiation emitted depends on temperature so the sphere emits brighter light when its temperature is greater ✓. The wavelength distribution of an emission depends on the temperature of the body ✓, so as the temperature of the sphere changed the wavelength of visible light it emitted changed, leading to a change in the colour from yellow to red ✓.

   The intensity and wavelength distribution of any radiation emitted depends on the temperature of the body.

For more on the topics covered in this chapter, see pages 30–41 of the *Collins GCSE AQA Physics Revision Guide*.

7 Magnetism and Electromagnetism

Permanent and Induced Magnetism, Magnetic Forces and Fields

This is a fairly straightforward topic area covering the key points of magnets and magnetic fields.

Example

An investigation was carried out into the properties of two **permanent magnets**. When one **pole** of one magnet was brought together with the pole of another magnet they **repelled** each other.

a) What statement could be made about these two poles? *(1 mark)*

> These poles must be like, e.g. north and north or south and south ✓.

The two poles on a magnet are north and south.

b) This is an example of a non-contact force.
Explain why. *(1 mark)*

> The magnets did not have to be touching in order to experience the force ✓.

Forces can be separated into contact forces and non-contact forces.

c) Strips of magnetic material were used in this investigation. Which of the following is not a magnetic material? Tick **one** box. *(1 mark)*

- iron ☐
- aluminium ☐
- cobalt ☐
- nickel ☐

> aluminium ✓

Four magnetic materials are listed in the specification – you should make sure you learn all four of these.

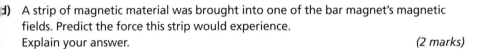
Magnetism and Electromagnetism

d) A strip of magnetic material was brought into one of the bar magnet's magnetic fields. Predict the force this strip would experience.
Explain your answer. *(2 marks)*

> It always experiences a force of attraction ✓, as the strip is an **induced magnet** and induced magnetism always causes a force of attraction ✓.

The strip is an induced magnet.

e) Whilst in the magnetic field of the bar magnet the strip could be used to pick up small magnetic objects. Explain what would happen to the strip's magnetism if it was removed from the bar magnet's magnetic field. *(1 mark)*

> The strip was an induced magnet so when removed from a magnetic field it loses its magnetism quickly ✓.

f) The magnetic strip was placed at different points around the bar magnet. At which part of the magnet would the strip experience the strongest force?
Explain your answer. *(2 marks)*

> The strip would feel the strongest force at the poles of the magnet ✓. This is the point where the magnetic field of the magnet is the strongest ✓.

The strongest force will be experienced in the area where the magnetic field of the magnet is the strongest.

g) As part of the investigation a compass was used to plot the magnetic field pattern around the bar magnet.
Explain how this could be carried out. *(3 marks)*

> Place the compass inside the magnetic field and the compass will point to the south pole of the magnet ✓. Draw a small line in the direction the compass is pointing ✓. Repeat for different points around the magnet ✓.

Magnetic field lines run from the north pole to the south pole of a magnet.

Magnetism and Electromagnetism

h) Explain why a compass can be used in navigation. *(2 marks)*

> A compass points in the direction of the Earth's magnetic field ✓. This means it will always point north ✓.

Use the fact that the Earth has a magnetic field in your answer.

i) What evidence does a compass provide about the core of the Earth? *(1 mark)*

> It provides evidence that the core of the Earth is magnetic and so generates a magnetic field ✓.

The Motor Effect

This is a more challenging topic, including applying knowledge of magnetic fields to practical applications of the **motor effect**.

Example

An investigation was carried out into the magnetic field produced by a **current** flowing through a conducting wire. A magnetic object was placed in the magnetic field.

a) Predict the effect of the following on the strength of the attraction experienced by the magnetic object.
In each case explain your answer.

   i) Increasing the distance from the wire. *(2 marks)*

   > Increasing the distance from the wire would decrease the force of attraction experienced ✓, as the strength of the magnetic field would decrease ✓.

   ii) Increasing the amplitude of the current flowing through the wire. *(2 marks)*

   > Increasing the amplitude of the current flowing through the wire would increase the force of attraction experienced ✓, as the strength of the magnetic field would increase ✓.

   Distance from the wire and the amplitude of the current flowing through the wire are two of the factors that affect the strength of a magnetic object placed in the magnetic field of an **electromagnet**.

Magnetism and Electromagnetism

b) What effect would shaping the wire to form a **solenoid** have on the magnetic field? *(1 mark)*

It would increase the strength of the magnetic field ✓.

> A solenoid is formed by wrapping the wire carrying the current into a coil.

c) A compass was used to plot the shape of the magnetic field around the solenoid. What statement could be made about the shape of the plot produced? *(1 mark)*

It would be the same shape as the magnetic field around a bar magnet ✓.

d) An iron core was added to the solenoid. What was now being produced? *(1 mark)*

An electromagnet ✓

e) The magnetic field was once again plotted. How would the plot have changed? Explain your answer. *(2 marks)*

The field lines would be closer together ✓, as the magnetic field would be stronger ✓.

> Placing an iron core into a solenoid increases the strength of the magnetic field.

 Example

During a school physics practical into the motor effect, a simple motor was constructed using a magnet and a wire.

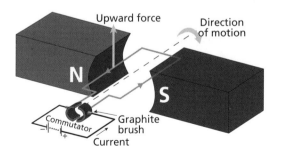

a) Explain what will happen when a current is passed through the wire. *(1 mark)*

The magnet and the conductor will exert a force on each other ✓.

Magnetism and Electromagnetism

b) Describe the effect of the following on the size of the force experienced by the conductor.

    i) Decreasing the strength of the magnetic field. *(1 mark)*

> Decreasing the strength of the magnetic field will decrease the force experienced by the conductor ✓.

    ii) Increasing the current flowing through the wire. *(1 mark)*

> Increasing the current flowing through the wire will increase the force experienced by the conductor ✓.

> The strength of the magnetic field and the current flowing through the wire are two factors which affect the size of the force.

c) What rule links the relative orientation of the force, the current in the conductor and the magnetic field? *(1 mark)*

> Fleming's left-hand rule ✓

In this investigation the magnetic **flux density** was 2.1 T, the current was 14 A and the conductor had a length of 10 cm. Use the correct equation from the Physics Equations Sheet.

d) What was the force experienced by the conductor? *(2 marks)*

> force = magnetic flux density × current × length
> force = 2.1 × 14 × 0.1 ✓
> force = 2.94 N ✓

> Convert the length of the conductor into metres before substituting the value into the equation.

e) The current was changed and a force of 1.89 N was experienced by the conductor. What is the new current? *(2 marks)*

> current = force ÷ (magnetic flux density × length)
> current = 1.89 ÷ (2.1 × 0.1) ✓
> current = 9 A ✓

> As the force has decreased the current must also have decreased.

Magnetism and Electromagnetism

Example

Use the diagram below to explain how a loudspeaker uses the motor effect to convert variations in current in electrical circuits to the pressure variations in sound waves. *(4 marks)*

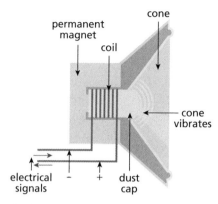

The electrical signal into the coil changes direction ✓, causing the electromagnet's magnetic field to change ✓. This causes the electromagnet to move back and forth ✓, vibrating the cone at different frequencies and generating pressure variations that result in a sound wave ✓.

Loudspeakers and headphones are specific examples of the motor effect, which you need to be able to explain.

Induced Potential and Transformers

This is a challenging topic area covering the **generator effect** and properties of transformers, including two key equations.

Example

a) Use terms from the box to complete the sentences below. Each term can be used once, more than once, or not at all. *(2 marks)*

| generator | current | resistance | potential difference | conductor | magnetic |

If an electrical _____ moves relative to a _____ field, a _____ is **induced** across the ends of the conductor. If the conductor is part of a complete circuit, a _____ is induced in the conductor. This is called the _____ effect.

Magnetism and Electromagnetism

As there are more words than gaps, at least one word will not be used.

If an electrical **conductor** moves relative to a **magnetic** field, a **potential difference** is induced across the ends of the conductor. If the conductor is part of a complete circuit, a **current** is induced in the conductor. This is called the **generator** effect. ✓ ✓

One incorrect answer –1 mark.

b) Explain how the following will affect the size of the potential difference produced by a generator.

  i) Increasing the speed of movement of the coil in the magnetic field.   *(1 mark)*

  Increasing the speed of movement of the coil in the magnetic field will increase the size of the potential difference produced by a generator ✓.

  ii) Decreasing the strength of the magnetic field around the coil.   *(1 mark)*

  Decreasing the strength of the magnetic field around the coil will decrease the size of the potential difference produced by a generator ✓.

  The speed of movement and the strength of the magnetic field are two factors which affect the size of the potential difference produced by a generator.

c) If the polarity of the magnetic field was reversed, what effect would this have on the induced potential difference?   *(1 mark)*

  The direction of the induced potential difference will be reversed ✓.

  The polarity of the magnetic field is the direction of the magnetic field, from north to south.

Magnetism and Electromagnetism

d) i) On the axis below draw an example of a current that is produced by a **dynamo**. *(1 mark)*

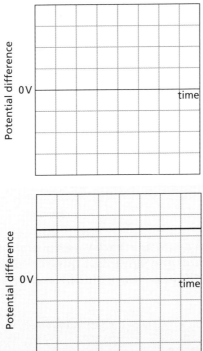

A dynamo generates d.c. current.

ii) Explain how the graph produced by an **alternator** would differ from the graph you have drawn. *(2 marks)*

This graph would show a.c. current as opposed to d.c. current ✓. This would mean that potential difference would increase then decrease, so the line would go up and down ✓.

You need to be able to draw and interpret graphs of potential difference generated in the coil against time for both types of generator.

Magnetism and Electromagnetism

Example

The diagram below shows the structure of a **microphone**. Explain, using the diagram, how the microphone converts sound waves into variations in current. *(4 marks)*

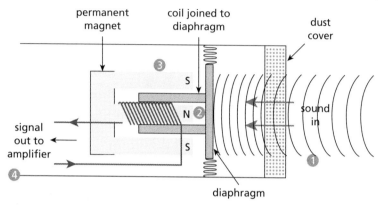

Sound waves move the diaphragm backwards and forwards ✓. The diaphragm moves the coil backwards and forwards ✓. This leads to a changing current induced in the coil ✓. The frequency of the current corresponds to the frequency of the sound ✓.

Microphones are a specific example of the generator effect, which you need to be able to explain.

Example

The diagram below shows the structure of a transformer.

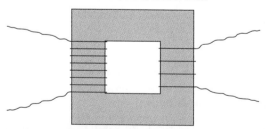

a) Explain why an iron core is used in the transformer. *(1 mark)*

Iron is used as it is easily magnetised ✓.

Iron is an example of a magnetic material.

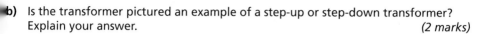

b) Is the transformer pictured an example of a step-up or step-down transformer? Explain your answer. *(2 marks)*

> This is a step-down transformer ✓, as it has more turns on the primary coil than the secondary coil. ✓

Determine if a transformer is a step-up transformer or a step-down transformer by counting the number of turns on the primary coil and the number of turns on the secondary coil.

c) The potential difference across the primary coil is 200 V.
What is the output potential difference?
Use the correct equation from the Physics Equations Sheet and show your working. *(3 marks)*

$$\frac{n_p}{n_s} = \frac{V_p}{V_s}$$

$$\frac{8}{4} = \frac{200}{V_s}$$

$$2 = \frac{200}{V_s} \checkmark$$

$$V_s = \frac{200}{2} \checkmark$$

$$V_s = 100 \text{ V} \checkmark$$

When answering this question, use the number of turns on the primary and secondary coils shown in the diagram.

d) The secondary coil on the transformer was modified so that a potential difference of 200 V across the primary coil produced a potential difference of 500 V across the secondary coil.
What was the number of turns on the secondary coil? *(3 marks)*

$$\frac{n_p}{n_s} = \frac{V_p}{V_s}$$

$$\frac{8}{n_s} = \frac{200}{500}$$

$$\frac{8}{n_s} = 0.4 \checkmark$$

$$n_s = \frac{8}{0.4} \checkmark$$

$$n_s = 20 \checkmark$$

As the potential difference across the secondary coil is greater than the potential difference across the primary coil, the transformer must have been modified to become a step-up transformer.

Magnetism and Electromagnetism

e) i) The current in the primary coil was 15 A.
If the transformer was 100% efficient, what is the current in the secondary coil?
Write down any equations you use. *(4 marks)*

> $V_s \times I_s = V_p \times I_p$ ✓
> $500 \times I_s = 200 \times 15$
> $500 \times I_s = 3000$ ✓
> $I_s = \frac{3000}{500}$ ✓
> $I_s = 6\,A$ ✓

The equation used to carry out this calculation can be found on the Physics Equations Sheet. The values for potential difference are given in part **d)**.

ii) In the example above an alternating current was fed into the primary coil. Would the transformer function if a d.c. current was passed through the primary coil?
Explain your answer. *(2 marks)*

> No, it wouldn't ✓.
>
> Transformers rely on the alternating current in one coil inducing a current in the other coil ✓.

d.c. stands for direct current.

For more on the topics covered in this chapter, see pages 66–71 of the *Collins GCSE AQA Physics Revision Guide*.

8 Space Physics

Solar System; Stability of Orbital Motions; Satellites

This topic focuses on the bodies in the solar system, with particular attention given to stars, and on the **orbital** motions of natural and artificial satellites.

Example

1) Use words from the box to complete the sentences below. Each word can be used once, more than once or not at all. *(2 marks)*

dwarf	satellites	star	planets

Within our solar system there is one _____, the Sun, plus the

eight _____ and the _____ planets that orbit

around the Sun. Natural _____, the moons that orbit planets,

are also part of the solar system.

> Within our solar system there is one **star**, the Sun, plus the eight **planets** and the **dwarf** planets that orbit around the Sun. Natural **satellites**, the moons that orbit planets, are also part of the solar system. ✓ ✓
> One incorrect answer −1 mark.

2) In the geocentric model the Earth is at the centre of the **Universe**, with the Sun, Moon, stars and planets all circling Earth. Many ancient civilisations held this model to be true. The evidence for the model is that from the Earth, the Sun appears to revolve around Earth once per day and the Moon and the stars also appear to revolve around Earth approximately once per day.

Explain which parts of this model are correct and which are incorrect and why a model of the Universe may change over time. *(4 marks)*

> The Moon does orbit around the Earth so this part of the model is correct ✓. The Sun does not orbit around the Earth, the Earth orbits the Sun, so this part of the model is incorrect ✓. The stars do not orbit the Earth so this part of the model is also incorrect ✓. Scientists gather new evidence and use this to change models ✓.

Space Physics

Example

The table below shows the diameter and mass of a star at different points in the star's life cycle. Stages A and B directly follow the star's **main sequence**.

Stage	Diameter (million km)	Mass (× 10^{30} kg)
Main sequence	1.4	1.97
A	200	1.59
B	0.014	1.19

a) What **two** forces are balanced in a main sequence star? *(2 marks)*

*The gravitational collapse ✓ and the expansion of a star due to **fusion** energy ✓.*

The Sun is currently in its main sequence.

b) The mass of the Sun is 1.99 × 10^{30} kg. What will be the final fate of this star? Explain how you arrived at your answer. *(2 marks)*

This star will form a black dwarf ✓ as it has a similar mass to the Sun.

A star's life cycle depends on whether it has a mass similar to the Sun or a larger mass than the Sun.

c) Identify the stages labelled A and B in the table.
Use the diameters of the different stages to explain how you arrived at your answer. *(4 marks)*

A is a red giant ✓, as its diameter is much larger than the star when it is in its main sequence ✓. B is a white dwarf ✓, as its diameter is much smaller than that of the main sequence star ✓.

d) The equation below shows a reaction that occurs in this star. It is part of a chain reaction.

$$^{2}_{1}H + ^{1}_{\square}H \rightarrow ^{\square}_{2}He + \gamma$$

Space Physics

i) What type of reaction is this? *(1 mark)*

   A fusion reaction ✓

ii) Complete the equation. *(2 marks)*

   Ensure the proton numbers and atomic numbers in the equation all balance.

   $$^{2}_{1}H + ^{1}_{1}H \rightarrow ^{3}_{2}He + \gamma$$ ✓✓

   One tick for each correct number.

iii) Explain why atoms of palladium would not be formed in this star. *(2 marks)*
   Refer to the Periodic Table to help with your answer.

   *Palladium is heavier than iron so is formed in a **supernova** ✓; this star will not undergo a supernova ✓.*

   Fusion processes in stars produce all of the naturally occurring elements. Elements heavier than iron are produced in a supernova.

Example

Divide the following objects into natural and artificial satellites. *(2 marks)*

a weather satellite   the Moon   Halley's comet   the International Space Station

Natural satellites	Artificial satellites

Natural satellites	Artificial satellites
the Moon	a weather satellite
Halley's comet	the International Space Station
✓	✓

One mark per column.

Space Physics

 Example

Ceres is the largest object in the asteroid belt that lies between Jupiter and Mars. Ceres has an orbital speed of 17.91 km/s. The orbital speed of Ceres remains relatively constant during its orbit around the Sun.

a) The orbital velocity of Ceres is constantly changing.
   Explain why. *(2 marks)*

> Speed is a scalar quantity so only has a magnitude, whilst velocity is a vector quantity, which has both a magnitude and a direction ✓. The velocity of Ceres is constantly changing as its direction is constantly changing during its orbit ✓.

> This question links back to the Forces topic and the difference between speed and velocity.

Ceres has a rotational speed of 92.61 m/s. If a future space mission put a satellite into geostationary orbit above Ceres, the satellite would have a speed of 92.61 m/s.

b) If the speed of the satellite changed, what would be the effect on its orbital radius to ensure the orbit remained stable? *(1 mark)*

> The orbital radius would have to change ✓.

Red-Shift

This topic focuses on **red-shift** as evidence for the Big Bang theory of the evolution of the Universe.

Example

The table below shows the red-shift of a number of different astronomical objects.

Name	Type	Red-shift
ULAS J1120 + 0641	black hole	7.09
CL J1001 + 0220	**galaxy** cluster	2.51
SN UDS10Wil	supernova	1.71
GN-z11	galaxy	11.09

Space Physics

a) What is meant by the term red-shift? *(3 marks)*

> There is an increase in the wavelength of light from the most distant galaxies ✓. The further away the galaxies, the faster they are moving and the bigger the increase in wavelength ✓, this leads to the light shifting to the red end of the spectrum ✓.

b) Which of the objects in the table is the most distant?
Explain how you arrived at your answer. *(1 mark)*

> GN-z11 is ths most distant as it has the highest red-shift, so must be the furthest from Earth ✓.

Use the red-shift values to work out this answer.

c) The Andromeda galaxy has a red-shift value of –0.001001. What conclusion can be drawn from this value? *(1 mark)*

> As this is a negative number it shows that the Andromeda galaxy is not moving away from the Earth but is moving closer to Earth ✓.

The key thing about this value is that it is negative – red-shift values are normally positive values.

The table below shows the wavelength of different colours of visible light.

Colour	Wavelength
blue	450–495 nm
green	495–570 nm
yellow	570–590 nm
orange	590–620 nm
red	620–750 nm

Space Physics

d) Use this information to explain why the Andromeda galaxy is said to be blueshifted. *(2 marks)*

> As Andromeda is moving towards Earth the wavelength of visible light is being decreased ✓. As blue light has a smaller wavelength, this means visible light is being shifted to the blue end of the spectrum ✓.

e) What galaxy is the Earth found in? *(1 mark)*

> The Milky Way ✓

f) Predict what will happen to this galaxy if the movement of Andromeda continues moving as it does now. *(1 mark)*

> Andromeda and the Milky Way will collide ✓.

Example

The Stationary Universe theory has the following characteristics:

- The Universe is an infinite size.
- The Universe is not expanding or contracting.
- The Universe has existed forever and will continue to exist forever.

This theory was first outlined in the 16th century.

a) Explain why the Stationary Universe theory is considered incorrect and what the most widely accepted theory of the start of the Universe is, including the evidence for this widely accepted theory. *(3 marks)*

> The Universe is expanding and is not an infinite size ✓. The Universe has not existed forever, it began with the Big Bang ✓. The evidence for this is provided by red-shift of galaxies moving away from Earth, with more distant galaxies moving faster ✓.
>
> Scientists use observations to arrive at theories such as the Big Bang.

Space Physics

The ATLAS detector at the Large Hadron Collider at CERN is being used by scientists trying to detect dark matter, an aspect of the Universe scientists still do not understand.

b) Give **one** other example of an aspect of the Universe that scientists do not understand. *(1 mark)*

> Dark energy ✓

Dark matter is one of the examples of things scientists don't understand about the Universe that is mentioned in the specification.

CERN research centre

For more on the topics covered in this chapter, see pages 94–97 of the *Collins GCSE AQA Physics Revision Guide*.

Glossary

Absorb / absorbed to take in and retain (all or some) incident radiated energy

Acceleration / accelerating the rate of change of velocity, measured in metres per second squared (m/s^2)

Activity the rate at which a radioactive source emits radiation, measured in becquerel

Alpha a type of radiation, which is strongly ionising, in the form of a particle consisting of two neutrons and two protons (a helium nucleus)

Alternating current (a.c.) a continuous electric current that periodically reverses direction, e.g. mains electricity in the UK

HT Alternator an electrical machine that uses a rotating magnet inside a fixed coil of wire to generate an alternating current

Amplitude the maximum displacement that any particle in a wave achieves from its undisturbed position, measured in metres (m)

Anomalous a data value that is significantly above or below the expected value / outside a pattern or trend

Atmospheric pressure the pressure exerted by the gaseous envelope surrounding the Earth or another body in space

Atom the smallest quantity of an element that can take part in a chemical reaction, consisting of a positively charged nucleus made up of protons and neutrons, surrounded by negatively charged electrons

Atomic number the number of protons in an atom of an element

Attraction a force by which one object attracts ('pulls') another, e.g. gravitational or electrostatic force

Beta a type of nuclear radiation that is moderately ionising, consisting of a high-speed electron, which is ejected from a nucleus as a neutron turns into a proton

Black body a body (object) that absorbs all the infrared radiation incident on it, i.e. does not reflect or transmit any infrared radiation incident on it

Braking distance the distance a vehicle travels under braking force (from the point when the brakes are first applied to the point when the vehicle comes to a complete stop)

Chain reaction a process in which a neutron colliding with an atomic nucleus causes fission and the ejection of one or more other neutrons, which induce other nuclei to split

Charge a property of matter that causes it to experience a force when placed in an electric field; electric current is the flow of charge; charge can be positive or negative and is measured in coulombs (C)

HT Collision an event in which two or more bodies, or particles, come together with a resulting change of direction and energy

Compression the act of squeezing / pressing (an elastic object)

Glossary

Concave having one or two surfaces curved inwards; a concave lens is sometimes called a diverging lens because parallel rays of light entering the lens spread out

Conductivity a measure of the ability of a substance to conduct electricity

Contact force a force that occurs between two objects that are in contact (touching), e.g. friction and tension

Contamination / contaminated the unwanted presence of materials containing radioactive atoms on other materials

Convex having one or two surfaces curved outwards; a convex lens is sometimes called a converging lens because parallel rays of light entering the lens are brought to a focus

Current the flow of electrical charge, measured in amperes (A)

Deceleration describes negative acceleration, i.e. when an object slows down

Dense having a high density (mass per volume)

Density a measure of mass per unit of volume, measured in kilograms per metre cubed (kg/m³)

Diode a component that only allows current to flow in one direction (has a very high resistance in the reverse direction)

Direct current (d.c.) a continuous electric current that flows in one direction only, without significant variation in magnitude, e.g. the current supplied by cells and batteries

HT Displace to cause a quantity of liquid, usually water, to move from its usual place

Displacement a vector quantity that describes how far and in what direction an object has travelled from its origin in a straight line

Distance a scalar quantity that provides a measure of how far an object has moved (without taking into account direction)

HT Dynamo a device that uses a rotating coil and a fixed magnet to convert mechanical energy into a direct current

Efficiency the ratio or percentage of useful energy out compared to total energy in for a system or device

Elastic potential energy the energy stored in a stretched / compressed elastic object, such as a spring

Electromagnet a magnet consisting of an iron or steel core wound with a coil of wire, through which a current is passed

Electromagnetic (EM) waves a continuous spectrum of waves formed by electric and magnetic fields, ranging from high frequency gamma rays to low frequency radio waves

Electron a subatomic particle, with a charge of –1, which orbits the nucleus of an atom

Element a substance that consists only of atoms with the same number of protons in their nuclei

Emit / emitted to give off (radiation or particles)

Energy a measure of the capacity of a body or system to do work

Extension the distance over which an object (such as a spring) has been extended / stretched

Glossary

Fission the splitting of an atomic nucleus into parts, either spontaneously or as a result of the impact of a particle, usually with an associated release of energy

Fluid a substance, such as a liquid or a gas, which can flow; has no fixed shape

Flux density a measure of the density of the field lines around a magnet

Focal length the distance from the centre point of a lens to the focus point, where the light rays converge / come together

Force an influence that occurs when two objects interact

Frequency the number of times that a wave / vibration repeats itself in a specified time period

Fusion a reaction in which two nuclei combine to form a nucleus with the release of energy

Galaxy a collection of millions / billions of stars held together by gravitational attraction

Gamma high frequency, short wavelength electromagnetic waves; a type of nuclear radiation, emitted from a nucleus

Gear a wheel with teeth that engages with another wheel with teeth, or with a rack, in order to change the speed or direction of transmitted motion

HT Generator effect to induce a potential difference (voltage) using magnetic fields and conductors

Gradient a measure of the steepness of a sloping line; the ratio of the change in vertical distance over the change in horizontal distance

Gravitational potential energy (GPE) the energy gained by raising an object above ground level (due to the force of gravity)

Gravity the force of attraction exerted by all masses on other masses, only noticeable with a large body, e.g. the Earth or Moon

Half-life the average time it takes for half the nuclei in a sample of radioactive isotope to decay; the time it takes for the count-rate / activity of a radioactive isotope to fall by 50% (halve)

Induce / induced to produce (a potential difference) or transmit (magnetism)

Induced magnet an object that becomes magnetic when placed in a magnetic field

Inelastically deformed describes an object that cannot return to its original shape when the forces that caused it to change shape are removed (because the limit of proportionality has been exceeded)

HT Inertia the tendency of a body to stay at rest or in uniform motion unless acted upon by an external force

Infrared the part of the electromagnetic spectrum with a longer wavelength than light but a shorter wavelength than radio waves

Internal energy the sum of the energy of all the particles that make up a system, i.e. the total kinetic and potential energy of all the particles added together

Ion formed when an atom loses or gains one or more electrons to become charged

Irradiation / irradiated to expose an object to nuclear radiation (the object does not become radioactive)

Isotope atoms of the same element that have different numbers of neutrons

Kinetic energy the energy of motion of an object, equal to the work it would do if brought to rest

Glossary

Latent heat of fusion the amount of heat energy needed for a specific amount of substance to change from solid to liquid

Latent heat of vaporisation the amount of heat energy needed for a specific amount of substance to change from liquid to gas

Lever a rigid bar set on a pivot, used to transfer a force to a load

Limit of proportionality the point up to which the extension of an elastic object is directly proportional to the applied force (once exceeded the relationship is no longer linear)

Longitudinal wave a wave in which the oscillations are parallel to the direction of energy transfer, e.g. sound waves

Magnification the ratio of image height to object height, e.g. magnification = image height / object height

Main sequence a stable period in the life cycle of a star, during which the outward acting fusion energy and inward acting force of gravity are in balance

Mass a measure of how much matter an object contains, measured in kilograms (kg)

Medium a material or substance

HT Microphone a device that uses the generator effect to convert sound waves into electrical signals

Microwaves electromagnetic radiation in the wavelength range 0.3 to 0.001 metres, used in satellite communication and cooking

Moment a measure of the turning effect of a force that causes an object to rotate about a pivot point, calculated by multiplying force by distance

HT Momentum the product of an object's mass and velocity

HT Motor effect the force experienced by a current carrying conductor when it is placed in a magnetic field, which is used to create movement in an electrical motor

National Grid the network of high voltage power lines and transformers that connects major power stations, businesses and homes

Neutron a neutral subatomic particle; a type of nuclear radiation, which can be emitted during radioactive decay

Non-contact force a force that occurs between two objects that are not in contact (not touching), e.g. gravitational and electrostatic forces

Normal at right angles to / perpendicular to

Nucleus the positively charged, dense region at the centre of an atom, made up of protons and neutrons, orbited by electrons

Ohmic conductor a resistor in which the current is directly proportional to the potential difference at a constant temperature

Opaque describes an object that either reflects or absorbs all light incident on its surface, so that no light passes through it

Orbit / orbital the curved path followed by a planet, satellite, comet, etc. as it travels around a body that exerts a gravitational force upon it; also applies to the paths of electrons around a nucleus (which exerts an electrostatic force of attraction)

Oscillate / oscillations to vibrate / swing from side to side with a regular frequency

Glossary

Parallel (circuit) a circuit in which the components are connected side by side on a separate branch / path, so that the current from the cell / battery splits with a portion going through each component

Particle an extremely small body with finite mass and negligible (insignificant) size, e.g. protons, neutrons and electrons

Period the time taken for a wave to complete one oscillation; the time it takes for a particle in the wave to move backwards and forwards once around its undisturbed position

Permanent magnet an object that produces its own magnetic field

Pivot the point around which an object turns

Pole the two opposite regions in a magnet, where the magnetic field is concentrated; can be north or south

Potential difference the difference in electric potential between two points in an electric field; the work that has to be done in transferring a unit of positive charge from one point to another, measured in volts (V)

Power a measure of the rate at which energy is transferred or work is done

Pressure the force exerted on a surface, e.g. by a gas on the walls of a container

Principal focus (also called *focal point*) the point where parallel rays of light travelling through a lens converge (meet) or from which they appear to diverge (spread out) from refraction by the lens

Proportional describes two variables that are related by a constant ratio

Proton a subatomic particle found in the nucleus of an atom, with an electrical charge of +1

P-waves (*Primary waves*) the longitudinal seismic waves produced during an earthquake

Radioactive containing a substance that gives out radiation

Real refers to an image produced by a lens, which is on the opposite side of the lens to the object and can be projected onto a screen (as opposed to a *virtual image*)

Red-shift the observed increase in the wavelength of light from distant galaxies, towards the red end of the spectrum

Reflected when a wave meets a boundary between two different materials and is bounced back

Refracted when a wave meets a boundary between two different materials and changes direction

Renewable can be replaced

Reproducible results are reproducible if the investigation / experiment can be repeated by another person, or by using different equipment / techniques, and the same results are obtained, demonstrating that the results are reliable

Repulsion / repelled a force that pushes two objects apart, such as the force between two like electric charges or magnetic poles

Resistance a measure of how a component resists (opposes) the flow of electrical charge, measured in ohms (Ω)

Resultant a single force that represents the overall effect of all the forces acting on an object

Glossary

Scalar a quantity, such as time or temperature, that has magnitude but no direction

Seismic caused by an earthquake

Series (circuit) a circuit in which the components are connected one after the other, so the same current flows through each component

Solenoid formed by coiling a wire to increase the strength of the magnetic field created by a current through the wire

Spark / sparking a momentary flash of light accompanied by a sharp crackling noise, produced by a sudden electrical discharge through the air between two points

Specific heat capacity the amount of energy required to raise the temperature of one kilogram of substance by one degree Celsius

Specular reflection reflection in a single direction (no scattering of light)

Speed a scalar measure of the distance travelled by an object in a unit of time, measured in metres per second (m/s)

Spring constant a measure of how easy it is to stretch or compress a spring; calculated as: force ÷ extension

State of matter the structure and form of a substance, i.e. gas, liquid or solid

Supernova a large star that explodes because the forces within it are no longer balanced / stable, releasing vast amounts of energy

S-waves (*Secondary waves*) the transverse seismic waves produced during an earthquake

System an object or group of objects

Tracer a radioactive isotope that is put into a system so that its movement can be tracked, helping to reveal blockages / holes that should not be there; used in medicine and industry

Transferred refers to how energy is changed, e.g. chemical energy can be transferred to electric energy

Transformer a device that transfers an alternating current from one circuit to another, with an increase (step-up transformer) or decrease (step-down transformer) of voltage

Transmitted when waves are sent out from a source or pass through a material

Transparent an object that transmits light coherently (the light rays do not get scattered), so that objects on the other side can be seen clearly

Transverse a wave in which the oscillations are at right angles to the direction of energy transfer, e.g. water waves

Ultrasonic / ultrasound sound waves with a frequency greater than 20 kHz, so they cannot be heard by humans

Ultraviolet the part of the electromagnetic spectrum with wavelengths shorter than light but longer than X-rays

Universe all existing matter, energy and space

Unstable lacking stability; having a very short lifetime; radioactive

HT Upthrust an upward push; the upwards force exerted by a fluid on an object in / partially in the fluid

Vector a variable quantity that has magnitude and direction

Glossary

Velocity a vector quantity that provides a measure for the speed of an object in a given direction

Virtual an image from which the light rays appear to come

Wavelength the distance from one point on a wave to the equivalent point on the next wave, measured in metres (m), represented by the symbol λ

Weight the vertical downwards force acting on an object due to gravity

Work the product of force and distance moved along the line of action of the force, when a force causes an object to move

X-rays the part of the electromagnetic spectrum with wavelengths shorter than that of ultraviolet radiation but longer than gamma rays

Formulae and Physics Equations Sheet

When solving quantitative problems, you should be able to recall and apply the following equations using standard SI units. This means these equations will **not** be given to you on the exam paper.

Word equation	Symbol equation
weight = mass × gravitational field strength (g)	$W = m\,g$
work done = force × distance (along the line of action of the force)	$W = F\,s$
force applied to a spring = spring constant × extension	$F = k\,e$
moment of a force = force × distance (normal to direction of force)	$M = F\,d$
pressure = $\dfrac{\text{force normal to a surface}}{\text{area of that surface}}$	$p = \dfrac{F}{A}$
distance travelled = speed × time	$s = v\,t$
acceleration = $\dfrac{\text{change in velocity}}{\text{time taken}}$	$a = \dfrac{\Delta v}{t}$
resultant force = mass × acceleration	$F = m\,a$
HT momentum = mass × velocity	$p = m\,v$
kinetic energy = 0.5 × mass × (speed)²	$E_k = \dfrac{1}{2} m v^2$
gravitational potential energy = mass × gravitational field strength (g) × height	$E_p = m\,g\,h$
power = $\dfrac{\text{energy transferred}}{\text{time}}$	$P = \dfrac{E}{t}$
power = $\dfrac{\text{work done}}{\text{time}}$	$P = \dfrac{W}{t}$
efficiency = $\dfrac{\text{useful output energy transfer}}{\text{total input energy transfer}}$	
efficiency = $\dfrac{\text{useful power output}}{\text{total power input}}$	
wave speed = frequency × wavelength	$v = f\,\lambda$
charge flow = current × time	$Q = I\,t$
potential difference = current × resistance	$V = I\,R$

Formulae and Physics Equations Sheet

Word equation	Symbol equation
power = potential difference × current	$P = VI$
power = (current)² × resistance	$P = I^2 R$
energy transferred = power × time	$E = Pt$
energy transferred = charge flow × potential difference	$E = QV$
density = mass / volume	$\rho = \dfrac{m}{V}$

You should be able to select and apply the following equations from the Physics Equations Sheet, which will be given to you as part of the exam paper.

Word equation	Symbol equation
HT pressure due to a column of liquid = height of column × density of liquid × gravitational field strength (g)	$p = h \rho g$
(final velocity)² − (initial velocity)² = 2 × acceleration × distance	$v^2 - u^2 = 2as$
HT force = change in momentum / time taken	$F = \dfrac{m \Delta v}{\Delta t}$
elastic potential energy = 0.5 × spring constant × (extension)²	$E_e = \dfrac{1}{2} k e^2$
change in thermal energy = mass × specific heat capacity × temperature change	$\Delta E = mc\Delta\theta$
period = 1 / frequency	
magnification = image height / object height	
HT force on a conductor (at right angles to a magnetic field) carrying a current = magnetic flux density × current × length	$F = BIl$
thermal energy for a change of state = mass × specific latent heat	$E = mL$
HT potential difference across primary coil / potential difference across secondary coil = number of turns in primary coil / number of turns in secondary coil	$\dfrac{V_p}{V_s} = \dfrac{n_p}{n_s}$
HT potential difference across primary coil × current in primary coil = potential difference across secondary coil × current in secondary coil	$V_p I_p = V_s I_s$
For gases: pressure × volume = constant	$pV =$ constant

Periodic Table

1	2												3	4	5	6	7	0 or 8
								1 **H** hydrogen 1										4 **He** helium 2
7 **Li** lithium 3	9 **Be** beryllium 4												11 **B** boron 5	12 **C** carbon 6	14 **N** nitrogen 7	16 **O** oxygen 8	19 **F** fluorine 9	20 **Ne** neon 10
23 **Na** sodium 11	24 **Mg** magnesium 12												27 **Al** aluminium 13	28 **Si** silicon 14	31 **P** phosphorus 15	32 **S** sulfur 16	35.5 **Cl** chlorine 17	40 **Ar** argon 18
39 **K** potassium 19	40 **Ca** calcium 20	45 **Sc** scandium 21	48 **Ti** titanium 22	51 **V** vanadium 23	52 **Cr** chromium 24	55 **Mn** manganese 25	56 **Fe** iron 26	59 **Co** cobalt 27	59 **Ni** nickel 28	63.5 **Cu** copper 29	65 **Zn** zinc 30	70 **Ga** gallium 31	73 **Ge** germanium 32	75 **As** arsenic 33	79 **Se** selenium 34	80 **Br** bromine 35	84 **Kr** krypton 36
85 **Rb** rubidium 37	88 **Sr** strontium 38	89 **Y** yttrium 39	91 **Zr** zirconium 40	93 **Nb** niobium 41	96 **Mo** molybdenum 42	[98] **Tc** technetium 43	101 **Ru** ruthenium 44	103 **Rh** rhodium 45	106 **Pd** palladium 46	108 **Ag** silver 47	112 **Cd** cadmium 48	115 **In** indium 49	119 **Sn** tin 50	122 **Sb** antimony 51	128 **Te** tellurium 52	127 **I** iodine 53	131 **Xe** xenon 54
133 **Cs** caesium 55	137 **Ba** barium 56	139 **La*** lanthanum 57	178 **Hf** hafnium 72	181 **Ta** tantalum 73	184 **W** tungsten 74	186 **Re** rhenium 75	190 **Os** osmium 76	192 **Ir** iridium 77	195 **Pt** platinum 78	197 **Au** gold 79	201 **Hg** mercury 80	204 **Tl** thallium 81	207 **Pb** lead 82	209 **Bi** bismuth 83	[209] **Po** polonium 84	[210] **At** astatine 85	[222] **Rn** radon 86
[223] **Fr** francium 87	[226] **Ra** radium 88	[227] **Ac*** actinium 89	[261] **Rf** rutherfordium 104	[262] **Db** dubnium 105	[266] **Sg** seaborgium 106	[264] **Bh** bohrium 107	[277] **Hs** hassium 108	[268] **Mt** meitnerium 109	[271] **Ds** darmstadtium 110	[272] **Rg** roentgenium 111	[285] **Cn** copernicium 112	[286] **Uut** ununtrium 113	[289] **Fl** flerovium 114	[289] **Uup** ununpentium 115	[293] **Lv** livermorium 116	[294] **Uus** ununseptium 117	[294] **Uuo** ununoctium 118

Key

- Metals
- Non-metals

Relative atomic mass → 1
Atomic symbol → **H**
Name → hydrogen
Atomic / proton number → 1

*The lanthanides (atomic numbers 58–71) and the actinides (atomic numbers 90–103) have been omitted.
The relative atomic masses of copper and chlorine have not been rounded to the nearest whole number.

Notes

Notes

Notes

Notes

Acknowledgements

The authors and publisher are grateful to the copyright holders for permission to use quoted materials and images.

Every effort has been made to trace copyright holders and obtain their permission for the use of copyright material. The authors and publisher will gladly receive information enabling them to rectify any error or omission in subsequent editions. All facts are correct at time of going to press.

Cover and p1 © iCultura Creative (RF) / Alamy Stock Photo, © Shutterstock.com
All other images are © Shutterstock.com and ©HarperCollins*Publishers*

Published by Collins
An imprint of HarperCollins*Publishers*
1 London Bridge Street
London SE1 9GF

ISBN: 978-0-00-827683-6

First published 2018
10 9 8 7 6 5 4 3 2 1
© HarperCollins*Publishers* Limited 2018

All rights reserved. No part of this publication may be reproduced, stored in a retrieval system, or transmitted, in any form or by any means, electronic, mechanical, photocopying, recording or otherwise, without the prior permission of Collins.

British Library Cataloguing in Publication Data.

A CIP record of this book is available from the British Library.

Commissioning Editor: Kerry Ferguson
Author: Dan Foulder
Project Editor: Charlotte Christensen
Project Manager and Editorial: Jill Laidlaw
Cover Design: Sarah Duxbury
Inside Concept Design: Paul Oates
Text Design and Layout: QBS Learning
Production: Natalia Rebow
Printed and bound in China by RR Donnelley APS